To Liz,
I hope this brings back
some great memories
of Ralph — and great
meeting you & working
with you on this book
signing!
Geoff Williams

INGENUITY *in a* CAN

INGENUITY *in a* CAN

the Ralph Stolle Story

GEOFF WILLIAMS

foreword by
CONGRESSMAN ROB PORTMAN

ORANGE FRAZER *PRESS*
Wilmington, Ohio

Book and jacket design by Jeff Fulwiler

For ordering information:

Orange Frazer Press

37 1/2 West Main Street, Wilmington, OH 45177

937.382.3196

www.orangefrazer.com

Library of Congress Cataloging-in-Publication Data
Williams, Geoffrey, 1970-
Ingenuity in a can : the Ralph Stolle story / by Geoffrey Williams.
p. cm.
Includes bibliograpical references and index.
ISBN 1-882203-45-3
1. Stolle, Ralph J. 1904-1996. 2. Inventors—United States—Biography.
3. Inventions—History. 4. Tin containers. I. Title.

T40.W53W55 2004
609.2-dc22

2004059958

ACKNOWLEDGEMENTS

by Geoff Williams

To Susan, for understanding that a writer's life is often not full of great wealth
To Isabelle and Lorelei, for making us rich beyond belief, anyway

WHEN I FIRST TOOK ON this book project, I didn't realize how stranded I would have been without the generous help of Ralph's family and friends who were willing to share their knowledge and anecdotes of one of the 20th century's underrated entrepreneurs. Without them, there would be no worthwhile book about Ralph Stolle, but without Cathy Chasteen, there would have been no book at all.

Cathy is one of Ralph's granddaughters, and she loved him very much. It's not just anybody who would go to a publisher and insist that they find a writer and publish a book about their grandfather. But Cathy did that, and she's been a pleasure to get to know. Her gentility and kindness are directly descended from her grandfather.

The people who knew Ralph loved him. If a falling safe had ever plunged from the sky, taking aim at Ralph, I have little doubt that his family, his friends, his employees, for Pete's sake, would have pushed him out of the way and taken the hit. That's the type of respect and adoration the man earned.

Thank you, too, to Ralph's friends, family members, and employees who were able to offer their insight: Dick Pope, former president of Ralph's company, was not only a walking encyclopedia on the Stolle Corporation, but he was also my first interview for the book, as well as one of the last, and he was informative and friendly every time I came over or called; George Henkle, who I visited three times, for a couple hours each, (along with his lovely wife, Lois), and who gave me some very helpful information about Ralph's early days, his middle days and his final years; Dr. Lee Beck, who became my new best friend during the last few weeks of writing this book, and who knew Ralph very well. Roger Aschenbach told me some great stories about Ralph's early days; Tom Anderson shared some vital information about the 1930s; Kathleen Stolle gave me crucial material about Ralph's last days; and Mary Jo Cropper, one of Ralph's daughters, showed me that Ralph wasn't all about business.

I'd also like to thank Harold Hall, Rita Hauck, Slim Buchanan (for driving me around Sidney and showing me where Ralph used to work, what hotel he occasionally stayed at, and where he got his hair cut.), and Ken and Ron Pendery, who were both very gracious with their time and family insight, as were Bob Mays and Nelson Schwab. I certainly appreciate Neil Armstrong's contribution, considering he took time out of his schedule from working on his autobiography, to email me some recollections of Ralph, and many thanks go to Elton Kaminiski, without whom Chapter Five would have had much less substance.

I'd also like to thank Campbell Graf, an early champion of this book and an all-around nice guy, Bob Yearout, for actually taping an interview with his mother, Vera Yearout, who knew Ralph way, way, way back when (and thank you, Vera, for your help), Bill Falknor, Ralph's grandchildren Spencer and Amy, and his daughters, Sandy and Gail, who gave me a lot of perspective. (That Mary Jo is in the book more than her sisters is because we did one extra interview, and she has email, which meant I could badger her with questions day and night.)

Ed Cranmer, one of Ralph's grandchildren, was indispensable, showing me around Ralph's ranch house and farm, offering me some interesting tidbits into Ralph's personality, and uncovering some papers of Ralph's from the 1920s dating to around 1940. Those papers—granted, mostly tax records—offered some crucial insight into Ralph's life back then.

Thank you to Maxine Pence, Richard Borchers, Dalton and Linda Messersmith, Art Middleton, Don Boyer, Roger Brautigam, Weldon Oakley, Norma Counts and Norman Jones, as well as the nice, helpful folks at the venerable Shelby County Historical Society, which is in Sidney, a city Ralph came to know well.

I also appreciate Ohio Congressman Rob Portman, who found the time to read this book and write the foreword during a time when he was constantly being pulled in a million different directions. It wasn't easy for him to find spare time.

And, of course, I want to thank my wife, Susan, and my young daughters, Isabelle and Lorelei. I'm particularly grateful to Susan, who didn't give me grief for working late a lot of nights writing this book, and who once made the observation that it's a good thing she became pregnant with Lorelei at the start of the project because it certainly wouldn't have happened during it.

Most of all, I want to thank Ralph Stolle, himself, for living such a rich and varied life that made it necessary to write a book about it, and for leaving just enough clues to follow so that I could write, what I believe, is an accurate and warm account of this very singular Ohio entrepreneur's ninety-two years on the planet. I regret that we never were able to meet in person. Maybe this will sound corny, but I'm going to miss him.

ACKNOWLEDGEMENTS

by Cathy Chasteen

Dedicated to Alexandra Catherine Lapp

IN MARCH OF 1998, my aunt Mary Jo Cropper suggested I meet with Campbell Graf to talk about the Ralph J. Stolle Free Enterprise Program that had been set up at Wilmington College after my grandfather's death. Campbell was serving as a consultant to the program. I did not know Campbell, but I knew his name. While I was a student at Wilmington College in the mid-eighties, my grandfather had insisted that I go meet Campbell, who at the time was the head of the co-op department at Wilmington College. I was to see Campbell, get to know him and tell him if there was anything I needed help with. I never made it to Campbell's office because I was a full time student with two small children. Not making time for Campbell was my mistake, because little did I know 12 years later I would meet him and come to consider him a close friend. We shared a bond. We both loved and appreciated my grandfather and we missed him terribly.

Challenges present themselves differently on a college campus than they do in the business world. Sometimes the cogs in the wheel move slowly and it is important to move in a forward direction. While the Ralph J. Stolle Free Enterprise Program was important, Campbell and I decided we had to take another road to preserve my grandfather's legacy. I called Campbell on the phone and said, "Campbell, it's time to write that book we have been talking about." Needless to say Campbell loved the idea. This book could not have been done if it wasn't for Campbell's excellent record keeping. He has note cards on every person he has ever met. One of those people happened to be Marcy Hawley, owner of Orange Frazer Press and publisher of this book. Marcy also liked the idea of a book about Ralph Stolle. She has been instrumental in the getting the project off the ground and seeing it through to completion. Marcy also deserves the credit for hiring such a wonderful, patient and articulate writer as Geoff Williams.

There were so many people to talk to and such a short time to do it in. And there was always the question of whether or not they would talk? My grandfather had always been so private and modest. Everyone close to him had respected that. To my surprise, the answer was a resounding yes! Many phone calls, visits and interviews ensued. One particularly memorable moment happened on a visit with George Henkle. Mr. Henkle put his arm around me and said "Cathy, this is the most important thing you are ever going to do." Needless to say, Mr. Henkle's words were a driving force for the project and they stayed with me throughout some sleepless nights.

I mentioned that my grandfather was a private man and I personally struggled with the question of whether he would approve of a book about his life. That struggle came to an end when I came across some post cards and a letter he sent to me from Thailand and Taiwan when I was barely six years old. The year was 1967. The postcards and letter spoke of the children from those countries and what their lives were like. They took me back to a time when I was six years old. I remember giggling and laughing and being so happy when I was with my grandparents. Kids know when they are loved and my grandfather and my grandmother loved us very much. A book was the least I could do to preserve those memories.

Ron and I have three children: Julia, Samantha and Joshua. They are fortunate enough to remember their great-grandfather. This book will insure they never forget his accomplishments and generous spirit.

And last, but certainly not least, is Alexandra Catherine, our granddaughter, the light of all our lives and the beginning of a brand new generation. Alexandra deserves to know what a special person her great-great-grandfather was and she deserves the opportunity to discover through the pages of this book where she came from and how far the American Dream can take her.

CONTENTS

FOREWORD

by Congressman Rob Portman

SHORTLY AFTER FINISHING Geoff Williams' account of the remarkable life of Ralph Stolle, his legacy seemed to come alive everywhere I went.

As we hustled through Blue Ash, Ohio, to one of the five Fourth of July parades I marched in this Independence Day, we passed the Stolle Milk Biologics building. As I drove my daughter to yet another soccer game, this one near Mason, Ohio, on a weekend afternoon, I glanced to my right and saw the Stolle Manufacturing Company. The parking lot was full with the cars of a generation of workers who may never have met the hard driving entrepreneur, but carry on his work. And, as I rushed to a lunch meeting at the Golden Lamb in Lebanon, Ohio, it would have been difficult not to notice one of the nation's largest YMCA's on the road into town, and the sign that bears Ralph Stolle's name and attests to his visionary approach.

This reference to Lebanon brings me to how I know Ralph Stolle, and why I am so honored to have been asked to pen a Foreword to this entertaining and comprehensive look at the life of this kid from Kentucky, who had such a positive influence on me, and on the area I represent in Congress.

During the period of Ralph Stolle's life when he was a traveling young businessman with farmland in Warren County, Ohio, he began to stop in at the Golden Lamb Inn in Lebanon for hearty meals and local gossip. The Golden Lamb was a stagecoach hotel which had been in business since 1803, the same year Ohio became a state. In 1926, long after the last stagecoach had passed, my grandfather, Robert H. Jones, became the proprietor. The once prominent four story brick hotel that had hosted ten U.S. Presidents and many other famous Americans of the 19th century had fallen into disrepair and demanded much of the young innkeeper.

My grandfather and Ralph had some things in common: they were about the same age, both were outsiders to Warren County, Ohio, and both were struggling young businessmen. The two of them became friends. Over the next 40 years, as Robert Jones and my grandmother, Virginia, lovingly restored the old inn to its previous prominence, that friendship deepened.

I was very close to my grandparents. My mother was an only child, and my brother, sister and I were showered with love and attention, stories and advice. My grandfather, and namesake, loved to tell me stories of people and places, and I soaked them up. One of the people he liked to talk about was his friend, Ralph Stolle.

He told me his own version of the stories in the book: Ralph's pioneering work with Alcoa and aluminum flip-top for cans, his innovative work with dairy cows whose milk could help fight disease (my grandparents and my mother used the Stolle powdered milk) and the Countryside YMCA vision, which my grandfather supported financially because of Ralph. But he also told me more. He told me of Ralph's Horatio Alger-type business successes, his work ethic and spirit of discovery. He also spoke with pride of the friendship between their families, and of Ralph's kindness.

These impressions of Ralph were reinforced by my own brief encounters with him and by my parents. My mother knew Ralph all through her life and, through her, my father, another small businessman and entrepreneur, got to know Ralph. In fact, he was one of those fortunate friends who attended the Stolle pheasant hunts at his farm. Their respect for his accomplishments and his character mirrored that of my grandparents and so many referenced in *Ingenuity In A Can: The Ralph Stolle Story*.

Now knowing even more about his extraordinary life story, I wish I had been able to spend more time with him prior to his death in 1996. I did have one interesting visit with Ralph shortly after I was elected to Congress in 1993. I had always been fascinated with his therapeutic powdered milk, so when he called one day to invite me to come see him at the farm and tour the dairy, I immediately accepted. I suppose it was the same experience many others had with Ralph, but it was special for me. He met me at the door of his farmhouse and office. I remember his gentle manner, but also the gleam in his eye as we surveyed his beloved farm through the large picture windows.

He spoke with conviction about the potential for curing disease and helping people live healthier lives through his powdered milk. His passion for the project was that of a man decades younger, or a visionary whose ideas kept him young. The dairy manager came by to take me on a tour through the barns, and I assumed Ralph, then 90 and walking with some difficulty, would wait for us to return. Not a chance. He insisted on marching through the barns and explaining the inoculation of the cow, processing the milk and his breeding program. Every detail.

I left that day feeling more energized and more optimistic. In person he had a way of inspiring you, and *Ingenuity In A Can* captures his essence, and thus does the same thing.

Thanks to this book, these days when I am traveling Ohio's Second Congressional District and come across his business or philanthropic legacies, and when I cross paths with the wonderful families of his three proud daughters, Sandy, Gail, and Mary Jo, or his wife, former judge Kathleen Porter Stolle, I will have a broader appreciation and a broader smile. For God smiled on our region and each of us when He sent Ralph Stolle into our midst.

INGENUITY *in a* CAN

Prologue

1963 PROL[OGUE]

"Never leave that till tomorrow which you can do today."

— BENJAMIN FRANKLIN, WHO COULD BE DESCRIBING RALPH STOLLE'S PHILOSOPHY

RALPH STOLLE ALWAYS LOOKED YOUNGER THAN HE WAS, EVIDENCED BY THIS PHOTO PROBABLY TAKEN DURING THE 1950S, WHEN HE WAS IN HIS 50S.

WINTER. 1963. It is well past midnight but not yet six. If we're going to be melodramatic, it's still a fair statement to say that even the President of the United States, John F. Kennedy, is sleeping. But another leader, albeit of a much smaller operation, in a modest-sized Midwestern city, is awake. This man slept many nights, of course, but this was a time when he more than occasionally patrolled Plant 6, listening to the hum of the machines and wanting to make certain that everything was going according to plan.

The man's name was Ralph Stolle, but everybody called him Ralph. He was about to change the world, but no one outside the walled fortress knew it. Even Ralph did not know it. Although he was attempting to create something with a lasting impact, he couldn't have guessed that when his time came, this would be—over thirty years later—the lead for the obituary writers in *USA Today* and *The New York Times.*

OGUE

Coke adds life. The Pepsi Generation. This Bud's for you.

All commercial slogans instantly recognizable to most of the public, but Ralph would have been hardly aware of them. He wasn't a man who watched much television, nor did he listen to the radio. He didn't go to the movies. He was middle-aged when Elvis hit his stride. Popular culture was for other people. That Ralph was about to begin a ripple effect across television and print advertising far into the future could not have crossed his mind.

But he was about to do just that. Ralph's contribution to the world would transform soda, the occasional small-canned juice manufacturer, and most definitely the beer industry. Even the tuna industry would ultimately clamor for what only his company could deliver: cans with an easy-open pull tab.

It is hard to imagine there was actually a day when you couldn't slide some coins into a vending machine and hear the satisfying clunk of a metal can tumbling down a chute. Every day, American television is inundated by commercials of beer-crazed friends doing anything for a Bud Light or a Coors, and more often than not, the people are gulping down liquid from a can. The biker in that commercial chasing a cheetah is upset that the animal has drunk his can of Mountain Dew. It is a silly point to make, that Ralph was indirectly responsible for the look and feel of that commercial.

The man's name was Ralph Stolle, but everybody called him Ralph. He was about to change the world, but no one outside the walled fortress knew it.

It's easy to be nostalgic for glass bottles—they had almost an artistic quality to them, and the television camera loved them. That was a Coca-Cola bottle that Mean Joe Greene chugged in his famous commercial that aired during the Super Bowl in 1980. But cans are safer than bottles—it's not a far-fetched scenario that on a hot summer barefooted day you drop a Coca-Cola bottle, step on a shard of its glass, and find yourself on the way to the emergency room. And there's no denying that it's the can of beer and especially soda pop that has become the container of convenience across the entire world, from Peoria to Pakistan and in every institution from boardroom meetings to college campuses.

In a world trying to save precious wasted minutes, the easy-open pull tab was a gadget begging for its own existence. Ralph didn't come up with the idea for the easy-open pull tab. That had already been done and patented. But until Ralph Stolle came along, no one had figured out how to mass-produce it. A good idea was yearning for its expression, but without the machinery to make it happen it was practically useless.

The invention was the easy-open pull tab, a tiny piece of metal taken for granted today. And no wonder. It's a doorknob to a can and it doesn't look like much: a simple piece of oblong-shaped aluminum with two holes in it. Nothing special, just some metal. But it was actually an engineering feat and one of the subjects in Stephen Van Dulken's history book, *Inventing the 20th Century: 100 Inventions That Shaped the World.*

Ralph had been down this road before. For the last forty years he had been improving upon inventions, but his work was the kind that rarely captures the imagination of the historians, journalists, or poets. For instance, Ralph's first business was an electroplating company. Several years later, he branched out—into anodized aluminum. Decades later, during the 1950s, Ralph and his key personnel were working on developing polystyrene foam insulation.

Funny, but Hollywood never came calling.

He made a quiet impact on just about everybody in the world without them even knowing it . . .

Ralph improved washing machines, dryers, and refrigerators. He had his hand in enhancing automobiles, aluminum siding, gutters, and roofs. But his companies—and he had many of them—also worked on equipment that the public is generally unaware of, such as industrial hydraulic and air cylinders, precision tools, dies, jigs, and fixtures. He once owned a business that mostly focused on manufacturing lift truck masts.

Yet this is a man who would create something that affected the lives of almost every person on the planet, both his as well as future generations: the billions upon billions of people who drink canned beverages.

This is also a man whose fingerprints are on the tower of the Empire State Building.

And he was a foot soldier in the battle to eradicate cancer.

Ralph was an entrepreneur before the word was fashionable. He made a quiet impact on just about everybody in the world without them even knowing it, except for one big thing, even if it looked like a little thing. And even then, as the beer industry made its debut with marvelous easy-to-open cans, the general public never knew it was largely because of a man named Ralph Stolle. And if they *had* known, they wouldn't have been able to pronounce it. He went a lifetime of having people mispronounce his name (it's Stah-lee, not Stole-lee).

All of this was fine with Ralph. He didn't want publicity. At least, that's what he said, and indeed, that's the way he acted.

During those winter nights in 1963, Ralph probably didn't whistle while he worked. He wasn't that sort, though he derived great pleasure from his career. The 59-year-old was a quiet type, gregarious when the occasion called for it, but mostly he played the role of a no-nonsense businessman. It was in his genes, and he was a

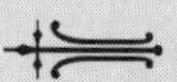

product of the era in which he grew up. He had learned early and often that the business world wasn't a game. This was a man who by the time he was 30 had battled a number of setbacks, including almost seeing his company go under because he was sick and battling a deadly disease. He was also one of millions almost crushed in the path of the Great Depression.

But he didn't merely survive; he thrived. Nothing, say the people who remember him, fazed Ralph Stolle. If business is your thing, and you want to understand leadership, people would say, this is the man to study.

Like most businessmen throughout most of the 20th century, Ralph always wore a tie and a jacket, and always a white, button-down shirt. It was his uniform. During the late nights, he probably forsook the tie. That he would be here after midnight made enough of an impression on the workers.

THESE CAN LID PROTOTYPES FEATURED AN EASY-OPEN PULL TAB, AN INVENTION THAT TODAY IS EASILY TAKEN FOR GRANTED.

What was going on during the winter of 1963 was the beginning of a mass-production marvel: seven million cans with the easy-open pull tabs were being produced a day. If they caught on with the rest of the world, many people throughout many industries would become rich. If the can lids were popular, a time-saving revolution would be launched. If they worked, it meant that aspiring entrepreneurs everywhere would have another role model from which to draw inspiration.

It *did* work, of course, as did just about everything else the man in the white shirt attempted. His name was Ralph Stolle, and this is his story.

Chapter One

1900s BEGINNI

"Entrepreneurs are simply those who understand that there is little difference between obstacle and opportunity and are able to turn both to their advantage."

— VICTOR KIAM, SUCCESSFUL ENTREPRENEUR WHO BECAME FAMOUS FOR OWNING REMINGTON ELECTRIC RAZORS

THE ALWAYS IMPECCABLY DRESSED RALPH J. STOLLE, AT HIS 1915 CONFIRMATION.

YOU DIDN'T SAY NO TO RALPH. That you would hop to whatever he asked was as preordained as the sun rising in the morning. A wildebeest would have better odds of convincing a leopard to stop smacking his lips and walk away. It wasn't that people were afraid of Ralph, although some surely were; they usually just accepted that he probably knew what was better for them, even more than *they* knew.

"He was always immaculately dressed, a good-looking man, who never showed his age, until he was really up there in years," recalls Tom Anderson, who first met Ralph in the 1930s. "He always looked like he had stepped out of a storefront window." Indeed, Ralph entered a room and heads swiveled; he commandeered a crowded room without trying, and he certainly didn't try, because he felt uncomfortable in crowds. People were naturally drawn to Ralph, in the way that a tree grows toward sunlight and waves embrace the shoreline. They couldn't help it. The man was a successful entrepreneur, generous, with an aura of invincibility about him. To not be a part of that would be like being invited to march in the Macy's Day Parade but opting to sort your laundry. You also didn't say no to Ralph because it would be like saying no to your father and disappointing him, and in later years, it would be like refusing the wishes of your grandfather. You said yes to Ralph because he was great and powerful, but

also because he was gentle and kind. There is no known story floating out there of an enraged Ralph losing his temper. He was King Kong without the roar.

He didn't start out great and powerful. His parents were strictly middle-class, sometimes a little upper, and sometimes a little lower. Frank Stolle made a decent living, but money was always in a short supply in a family with six children. And Ralph was the first to arrive in the household of Frank and Kunigunde, a woman born in Kentucky to Reinhard Pierman and née Charlotte Margaretha Dorner.

Ralph came into the world—in Newport, Kentucky, specifically—on January 17, 1904. He was born Ralph Johann Stolle, though later he would Americanize his middle name into John, possibly during the anti-German sentiment that swept the country during World War I. It was a different world than the one in which his parents had been born—Kunigunde, on October 18, 1880 and Frank, July 25, 1878. Icons from the 19th century were aging and fading. A white-haired Mark Twain would only last another six years. The soldiers of the Civil War, once decked in blue and gray, were now all gray. The Indians that once roamed free were killing time on reservations.

Meanwhile, the 20th century that Ralph was born into was just getting started. Bob Hope was in diapers; Cary Grant would soon join him. Henry Ford was just a year into his new automobile company. In April of 1904, the St. Louis World's Fair introduced the hamburger to the planet, as well as the hot dog, iced tea, and the ice cream cone. In the same city, police battled the elements by trying to identify criminals with a newfangled idea—fingerprinting. The first subway tracks were being laid in New York City, and less than 100 miles from the Stolle household, the airplane industry was getting off the ground in Dayton, Ohio. Before the year was up, the Wright brothers would successfully remain in the air for more than five minutes.

Everyone who knew Frank and Kunigunde remembers a household where there was one clear leader, and one follower. (Hint: the leader wasn't Frank.)

But some of Ralph's earliest memories have nothing to do with history, rather they belong somewhere in a Stephen King novel: the cracking of ice on a frozen pond near Grant Street, where his grandparents lived; a young cousin plunging into the icy water and drowning. All Ralph could do was watch helplessly. Logically, there was little or nothing that Ralph could have done. He was young when the tragedy struck, perhaps around 10 years old, and even a grown-up, who would have fared no better on the ice, would have been hard pressed to help. As far as anybody alive recalls, Ralph never spoke about the incident, and the name of the cousin remains a mystery, but fragments of the story have remained in the family.

From that day forward, Ralph considered the water his nemesis. He never learned to swim but insisted that his three daughters all take lessons, and he demanded that the grandchildren follow suit, too. And early in his fatherhood, in the 1930s, when Ralph and his wife moved to Carolina Street, their house had a swimming pool in the backyard. Ralph filled the swimming pool with dirt, and topped it off with grass seed. After the blades came in, and the grass waved in the wind, it was just an indistinguishable part of the back yard.

There was one other way Ralph internalized the incident. He learned that life was to be savored, not squandered. Ralph never wasted a minute, achieving as much in one lifetime as it might take a hundred people to do. He didn't sip from the cup of life; he drank in gulps.

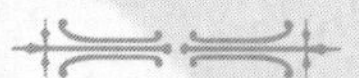

Everyone who knew Frank and Kunigunde remembers a household where there was one clear leader, and one follower. (Hint: the leader wasn't Frank.) Despite it being an age where the father was firmly in charge of the household, this wasn't the case in the Stolle residence. If Kunigunde said jump, Frank did. "She was the commander-in-chief," says Ken Pendery.

"She ran the family," adds his brother, Ron. "But he accepted her dominance. He knew when to speak, and when he could get away with what he could get away with, and how to do it."

Kunigunde was a force of nature, a gust of wind, impossible to ignore. She was the sort of woman who years later played cards with her sons into the wee hours of the morning, and chugged down a few beers. Yet at the same time, she could be a gentle breeze, warm and playful, content to dote on her grandchildren.

But during her early adulthood, Kunigunde didn't have the time to spend hours playing cards or with the children, not with so much to do around the house. Meanwhile, Frank was often across the Ohio River in Cincinnati, working as a sales representative for a distributor that manufactured steel bars. It was a job he started around the time he married Kunigunde; before that, he toiled in a machine shop in Cincinnati, commuting back and forth every day in a horse and buggy.

On the homefront, Kunigunde encouraged her children to make money themselves, by selling eggs from the chickens, vegetables from the garden, or breeding rabbits. Kunigunde especially pressed Ralph and Howard to turn whatever they could into a money-making opportunity. Ralph took to it, developing it into an art form. There's one rather gruesome story of how Ralph as an adult was playing first base on a baseball team for one of his company's teams, and he broke his finger. His company's health insurance paid for medical bills, but instead of using the money to fix his finger, Ralph did that himself and invested the money in stock.

When the stock paid dividends, he put it back into the stock. Fifty years later, when the company was sold, Ralph had to sell his stock, and he received a million dollars. This is what Kunigunde had wrought.

Kunigunde wasn't only fueling Ralph with a desire for making money. She was shaping him in many ways—ways that she could possibly not have dreamed at the time—because as the 20th century aged into its teens and Kunigunde entered her 40s, she was more distracted with the entrepreneur skills of her *husband.* After all, as the children grew, so did their expenses and needs. Not to mention that coming from such a big family, there were occasionally other people shuttling in and out of the house. The 1910 census shows that Frank's 21-year-old sister, Anna, was living with them, and sometime later, Vera Schwitzer, a cousin of Ralph's from his mother's side, moved in with them. Fourteen years old and an orphan, Vera lived with Frank and Kunigunde until going to nursing school.

FRANK AND KUNIGUNDE STOLLE, AT THE TURN OF THE CENTURY, ON THEIR WEDDING DAY.

The pressures of having to feed, clothe and shelter so many young people helps explain why every night at the dinner table, Kunigunde grilled Frank. She was the chicken, and he was the feed: Frank was henpecked. As Ralph, Howard, Lester, Margaret, Charlotte, and Irma looked on, Kunigunde wanted to know how Frank's day had gone, what leads he had chased down, what deals he had secured and, ultimately, how much money would soon be coming into the Stolle household. If Frank didn't mind the evening inquisitions, Ralph did.

Ralph vowed that this wasn't going to be the model for dinner conversations in his future household. Years later, when Ralph was married, he kept his business at the business. His spouse knew virtually nothing about the companies he cultivated from the 1920s and would cultivate well into the 1990s. Years later, when he sold part of his company for a substantial amount of money, he shared what he had just earned with his wife and instead of congratulating him she instead worried about how they would be taxed. Ralph was amused but also a little hurt. When it came to

wives and business, he felt he couldn't win. It was old-fashioned thinking, but he had reason for it, and he never managed to shake his feeling that the two factions should never mix. Deep into the 20th century, when Ralph was older and his gait was slow—forget that women had long since been voting and working, even fighting in the military—he advised one of his grandsons, Ed Cranmer, to never share the business side of life with his wife.

But Kunigunde was teaching Ralph lessons other than the value of making a dollar. While Kunigunde was in charge of the household, and Frank was in second place, Ralph was the third-in-command. During the summer months, when Frank was at work and the kids were out of school, Kunigunde often kept Ralph from playing outdoors—training him, instead, to help take care of the other children. He disliked being a secondary parent to his siblings, and he resented his mother for it, but there was no denying that she was shaping his personality into one that would serve him well. Ralph was learning to be a boss.

For starters, Ralph didn't always appreciate being bossed around, and he developed an authoritative style that would not be quite as dictatorial as his mother's, but just as effective. His lifelong pal, George Henkle, recalls hearing Ralph say, "My mother was like an army sergeant. My mother was tough." And Ralph told Henkle about being sent out with the other children to clean up the yard after a storm. They diligently cleared away branches and trash, washed walls and windows and restored the barn back to its luster. Once everything seemed in tip-top shape, they reported back to Kunigunde, who wasn't impressed and sent them to do everything all over again.

When Ralph later became an employer . . . his gift was for making people want to do things for him, instead of having them feel bullied into it.

The chores were a way of life. Ralph would talk of waking up in the early morning, and before school collecting eggs from the chickens and milking two cows, all of which he enjoyed. In his later years, Ralph recalled during a rare interview, "I milked two cows from the time I was six years old until I got out of high school. And those times would be at midnight and all hours of the day, and from then on, I always longed to have a farm." One of Ralph's chores was shoveling manure, which he *didn't* enjoy nearly as much.

"She was ungodly tough on those kids," says Henkle. And it doesn't take much imagination to recognize that Ralph didn't want his employees to resent him, in the way that he sometimes had felt about his mother. When Ralph later became an employer, it was clear that he wanted things done a certain way, but his gift was for making people want to do things for him, instead of having them feel bullied into it.

And Ralph wasn't just learning to be a chief but a chef. Had Ralph not gone into the metal industry, it isn't difficult to imagine him having opened up a restaurant or started one of the first sit-down restaurant chains. With six children in the house, there were plenty of plates to fill, and so with Kunigunde as his mentor, part of Ralph's duties was to cook in unbelievably large quantities.

Decades later, as a business leader, one of Ralph's most distinctive qualities was that he would play the role of chef for his associates in the corporate world. If you were a bigwig from a multi-million dollar firm, or the lowliest janitor in the Stolle business, it didn't really matter: At some point, probabilities were, Ralph would end up preparing a meal for you. And it charmed everybody that this millionaire entrepreneur would take the time to do that.

But some of the cuisine that young Ralph was expected to fix for the family wasn't for the squeamish. Ed remembers his grandfather talking about making "blood pudding." Ralph was sent out to butcher a hog and afterward the animal was bled into a skillet, and that was fried up as a meal before the family took to eating the other parts of the pig. Small wonder that Ralph grew up to be somebody you wouldn't want to meet in a dark corporate alley. His upbringing, not to mention genetics, practically demanded that he harden himself up. Even though Ralph was unfailingly pleasant, to the point that his employees and friends wanted to swim oceans and climb mountain peaks for him, there was always that idea in the back of your mind that if you were to cross Ralph and do him wrong, you would regret it.

Which was sometimes true, and sometimes not. People did disagree with Ralph and live to tell the tale. But Ralph gave that aura of strength and rugged character that one might find in a John Wayne. But there were exceptions. For instance, Ralph was petrified of public speaking. Even in somewhat intimate settings with a dozen or so people, the Stolle Corporation's leader was curiously mute. He usually said nothing, other than a few mumbled words, until the meeting was over and people were filing into the hallway, and then he would draw a person aside and tell him what he actually thought of an idea.

Ralph never met his great-great grandfather, but the two would have gotten along well. For Ralph, just beginning a business was a great risk; for 30-year-old Joseph Stolle, *his* great risk was bringing his wife and possibly some young children to America in 1836 from Lichtenberg, Alsace, which used to be in Germany but now is a region in northeastern France. Joseph, his wife, née Catherine Durr, traveled cross-country and stayed in Zanesville, Ohio, for a spell before settling in a tiny town called White Oak, now a suburb of Cincinnati. Exactly what Joseph did for his livelihood may be lost to time, but he did have a farm and contained some of the entrepreneurial spirit that would be passed down to Ralph; he owned 160 acres of land upon his death. Joseph Stolle was a tall man, with big hands, black and thick eyebrows, with white hair and a long goatee hanging off his chin. In what may be the only photo of him to exist— excepting the beard—he looks a little like Gregory Peck.

As early as the middle of the 19th century, Joseph's son, Frank, was also bitten by the business bug: He dabbled in real estate, when he wasn't working for himself as a tailor, which surely came in handy when clothing his seven children.

One of those children was named John, born on August 27, 1855, approximately two years after his grandfather, Joseph, passed away. John would ultimately have seven children, one of whom was named Frank, who in turn had six children, one of whom was named Ralph.

When Ralph escaped the clutches of his mother, he was usually at his grandparents' house, following around his grandmother, Anna, a slender woman with her hair pulled back in a bun. His grandfather, John, also had been too active over the years to gain any significant weight; he was a wiry man who had a heavy beard until he replaced it with a walrus moustache. Both seemed to be in fine health; Anna lived to see her grandson reach his 20s, and John died the same year Ralph turned forty.

The demands were less here with his grandparents, and he could escape the crowded house. It was only John, Anna, and probably one of their children, now a full-fledged adult, also named Anna. She never married and may have had emotional or mental challenges, according to Ken and Ron Pendery, who both have memories of seeing her at John Stolle's house in the 1930s and early 1940s.

More than his grandfather John, it was Anna, Sr., with whom Ralph developed a close bond. They spent a lot of time together. Although there was no genetic link between Anna Stolle and Kunigunde, the elder Mrs. Stolle was no pushover either. Ralph later described his grandmother as a tiny, tough woman, who would hit you with a wooden spoon if you got out of line. But the two clearly loved each other. Ralph spent every minute of free time that he could with his grandparents, and he particularly loved milking their cow. It was from his grandmother that he became interested in farming, an avocation that later consumed him.

There were other influences as well. Ralph's father wasn't an entrepreneur but numerous relatives were. Continuing the Stolle streak for owning property, John and Anna had spent their younger days operating a popular tavern in Northern Kentucky called Stolle Grove. John had worked hard throughout the years; for a brief time, he commuted to Cincinnati via horse and buggy to work as a clothing cutter, which is a similar career to his father's tailoring job but different in that one marked and cut fabric from a pattern. John Stolle was also a truck farmer. It was a career that had nothing to do with trucks; farmers trucked produce by horse and buggy to local markets to sell them. And so Ralph grew up, watching his father work for somebody else and being called to task at the dinner table for it, and his grandfather working for himself and having only himself to answer to. Ralph also watched his parents struggle to pay the mortgage and taxes. Even if it wasn't a fair comparison—John and Anna were further ahead in life than Frank and Kunigunde, and John and Anna's kids were mostly out of the house—there was no doubt who had better lives.

Meanwhile, John's brother—Ralph's great-uncle—was also an entrepreneur, arguably the most successful Stolle yet. Anton Stolle had taken his brood to Richmond, Indiana, and founded a meat packing company known as Stolle & Sons.

"Half the time I was playing basketball," Reuben would later tell his children, "and the other half, I was picking the splinters out of Ralph Stolle's butt."

But during his teenage years, Ralph wasn't really thinking about running his own company. If anything, he was more consumed with trying to find the time and resources to help put food on the table and still attempt to be a normal kid. One way he did both came from his high school drafting class. The teenager impressed his teacher so much that the teacher hooked the boy up with a building contractor, and Ralph wound up designing several houses in Ft. Thomas. It also brought some extra money to the Stolle household, which didn't hurt.

That Ralph could sometimes just be a kid and not think about earning a buck was in large part thanks to his pal, Reuben Aschenbach, a boy he had met at his church. Reuben was four years older, but the two had a rhythm going that kept them friends to the ends of their lives. During the beginning of their friendship, they got to know each other by meeting in church—Ralph was a devout Presbyterian—and by playing on the same church basketball team, dribbling, scoring, and tripping on rough wooden floorboards. "Half the time I was playing basketball," Reuben would later tell his children, "and the other half, I was picking the splinters out of Ralph Stolle's butt."

For sustenance—a growing boy must eat—Ralph and Reuben enjoyed sneaking past the house of a neighbor, Daisy Schmidt, and stealing the pies off her windowsill, where she placed them to cool. Another story Reuben liked to tell was of the time he and Ralph were recruited to move furniture into their minister's house. He had left town to get married, and Ralph and Reuben helped several other

churchgoers move the furniture in and get the house ready. Somehow, Reuben told his son, Roger, a basement window was left unlocked. And later that night, Reuben and Ralph snuck into the minister's house, gleefully putting living room furniture into the bedroom, the bedroom furniture into the kitchen, and so on. "And I think they got away with that," says Roger.

Ralph also had plenty of chances to be a kid in high school, where, during the after hours, he played baseball, basketball, tennis, and he ran track, specializing in the 220-yard race and the 440. During his senior year, he also participated in the class play. What his role was in the play seems to have been lost to the ages, but it seems he was a hit. In his 1921 yearbook, friends predicted that Ralph Stolle and a classmate, Frank Stegeman, would have a future "in the sideshow of a circus" and another prediction had Ralph penning a play or one-act, "Will Dorothea be My Aunt?" That last prediction had nothing to do with acting, however, and it would come true. His comely classmate, Dorothea Macht, ended up marrying an uncle of Ralph's, who was older by only five years, and did become his aunt.

"Ralph was never much of a ladies' man, but if he gets started, boys, clear off the stage."

If his yearbook is any indication, Ralph was well-liked, at least by the people writing it, and despite having a tough year with some classmates, he had some fun. He was a member of a Boys' Club at one point, and significantly, during his senior year, he was the vice-president of the Hi-Y Club.

The Hi-Y Club was an after-school organization originally developed by the YMCA in the 19th century and was extremely popular during the first half of the 20th century, though some Hi-Y Clubs still exist today. At the time, it was a group only for the high school boys—America's first Hi-Y club for girls wouldn't exist until two years after Ralph graduated, in 1923. If he had been around girls in the Hi-Y Club, it might have helped Ralph. As a teenage boy, Ralph was always painfully shy around girls, something his wife would fondly recall to friends years later, and something his high school buddies ribbed him about. In his 1921 yearbook, somebody came up with the following tribute: "Ralph was never much of a ladies' man, but if he gets started, boys, clear off the stage." And another comment in the yearbook jokes that Ralph was often seen with "the girls." If that was true, the writer was referring to his three sisters.

But while the Hi-Y Club wasn't offering Ralph much of a chance to improve his social skills with the fairer sex, the organization was at least burnishing his talent for talking to people, one-on-one, which he would use later in his business. According to an official YMCA historical document on the subject, the Hi-Y

Clubs offered a structure that allowed kids to plan their own activities, like sports, debates, and corn roasts. It all made a huge impression on Ralph, who would later believe that most of the problems of deviant youth could be solved, if only there was a YMCA program in which they could get involved. Ralph never forget his association with the Hi-Y Club, and the carefree days it offered him.

Those carefree days began skidding to a stop shortly before Ralph's senior year in high school. He was an ambitious student, which led to his departure from high school after just three years. Ralph was 17 and not happy. In one of the few documents that he wrote and still exists, probably typed up in the 1960s, Ralph penned a "history of Ralph J. Stolle." He leads off the two-page history with, "I had the misfortune of graduating from High School in three years because you had to graduate when you had so many credits...."

Ralph discovered this as his second year came to a close, which necessitated him and several other classmates to be moved up a year and take senior classes. "Usually your final year in High School is your best year, which I missed," wrote Ralph. Sandwiched into a new grade, Ralph reported that his new classmates weren't very friendly to him, increasing his anxiety over cutting his education career short. It is suspected that he was ostracized by some of the wealthier teenagers, who didn't know Ralph as well as their other classmates. "Fort Thomas had some very wealthy families, and he never felt like his family fit in," observes his daughter Mary Jo.

It didn't help matters that every day, Ralph walked to and from school, passing an especially affluent home. *Someday, I'm going to live there,* thought Ralph, who never wound up there but did well for himself anyway. Still, in high school, the house and the high society families were all daily reminders of what he didn't have. It was an insecurity that Ralph would never quite shake. After his death some seventy years later, the family discovered that the pantry was packed with food. Spencer Cropper, Mary Jo's son and a grandson of Ralph, describes it as being "a mini-Kroger," referring, of course, to the nationwide grocery store and conglomerate. All of the food that Ralph kept around "had to be because of a fear of running out of something," theorizes Mary Jo.

Still, in high school, the house and the high society families were all daily reminders of what he didn't have. It was an insecurity that Ralph would never quite shake.

But with college right around the corner, more prosperous times would surely be incubated. Ralph had a plan mapped out. He would attend the University of Cincinnati as a chemical engineering student. It was an astute decision. During the 19th century, the science of chemistry truly began to be understood, and during the 20th, people were figuring out how to make money from it. As Ralph began his freshman year of high school, for instance, just one state away in Terre Haute, Indiana, chemical engineers had developed acetone, a colorless liquid that would be used to make plastic, fibers, drugs, and other chemicals. Scientists in the 1920s would discover how to make cellulose acetate and acrylics in large quantities, bringing us Lucite and Plexiglass. During the same decade, meanwhile, universities across the world were adding chemical engineering departments. The University of Cincinnati, however, had had a chemical department since 1900, just twelve years after the first chemical engineering curriculum had been developed at MIT.

This was the world of chemical engineering in which Ralph lived. But reality quickly smacked him in the face. "Unfortunately, my Dad went financially broke in an investment, which meant that I had to find a job," Ralph would write, years later. "I got a job in a warehouse cutting steel. It was necessary that I get to the warehouse at 4 o'clock in the morning and load up saws that would cut disks for making gears."

If he were going to jettison himself from this grueling lifestyle . . . working for somebody else was not the answer. He would have to start his own company, and soon. Even if he was only 17.

But Ralph didn't abandon the idea of getting an education. In the same document that he wrote during the 1960s, Ralph described attending the University of Cincinnati as a part-time student, visiting the campus in the morning, studying engineering and chemistry, returning to the warehouse by noon, then going to the Ohio Mechanics Institute in the late afternoon. Friends remember hearing that Ralph attended night school, so it's quite possible that his classes took him well into the evenings or he eventually abandoned day classes for night ones. He needed, of course, to cram some time in his day for studying, yet manage to get to bed at a reasonable hour so he could wake up at 4 a.m. and repeat the entire schedule.

Ralph was aware this couldn't go on. It didn't help the teenager's morale that just as this financial crisis had hammered his family, the nation as a whole was coming out of a brief post-war depression and heading into the Roaring 20s, a time of unprecedented heights of prosperity. It was 1921, and young couples were flocking to see Rudolph Valentino in *The Sheik*. Women were letting down their hair—

which was more difficult since it was becoming shorter and shorter. Known as flappers, they were celebrating everything from new creations like Chanel No. 5 to their new-found right to vote. The Jazz Age was heating up. It was 1921, and people were making money and having fun as they welcomed in a new era—while Ralph was cutting his college life short and cutting steel in a stifling, humid warehouse.

It isn't difficult to imagine what inspired Ralph to make the most important decision of his life. If he were going to jettison himself from this grueling lifestyle and no longer be dependent upon his parents, working for somebody else was not the answer. He would have to start his own company, and soon. Even if he was only 17.

Chapter Two

1920s ILL WINDS

"The chief business of the American people is business."

— CALVIN COOLIDGE, IN A SPEECH ON JANUARY 17, 1925, RALPH'S 21ST BIRTHDAY

LEAVE HOOFING IT FOR THE HORSES: RALPH TOOK THE STREET CAR TO GET TO HIS NIGHT CLASSES.

THE CALENDAR WOULD TURN another year older, before Ralph Stolle would start his own company.

It was 1922, when Ralph, 18 and desperate to get his adult life going into a meaningful direction, read an ad in a newspaper, which would jumpstart his future. After the passages of time, what happened next is a little fuzzy but what we know goes something like this. Ralph paid $250 to that owner in the ad, to purchase the operation whole. Fifty dollars, it is said, Ralph had earned himself; the other $200, it's believed, came from his beloved grandmother, Anna. And so for the first time, Ralph mixed business with family. It wouldn't be his last.

Ralph set to work. What his $250 bought were tanks and batteries, which he immediately set up in the family garage. His parents, Ralph must have decided, could park on the street.

It goes without saying that electroplating can't go without saying. It's not an industry widely understood, or noticed, by the general public, and yet if it disappeared from the planet tomorrow, the world would *definitely* notice. And that was one of Ralph's hallmarks, to engage in a process with which no one is familiar, and to create a product with which *everyone* is familiar.

AHEAD

Electroplating began in the early 1800s, though it really took off in the 1850s. It is the method of turning nickel, brass, tin, zinc, and other metals into a coating, coatings that can be used on everything from streetlights to tin whistles and sink faucets. There's a real science to it, the kind of technological wizardry that makes the eyes of English majors glaze over. But a metal coating isn't created by magic; you need to put a negative charge onto the object, sink it into a special solution, and hope that the metallic ions of that solution both carry a positive charge and are attracted to the object you're trying to coat.

It sounds confusing to the average layman, but Ralph understood it. Possibly it was a trade he had picked up in the last year cutting steel at the warehouse, and it probably helped that his father was in the steel industry. In any case, Ralph understood that being able to electroplate the products other businesses were making would make a fine business for himself. And so he began approaching local companies and convincing them to let the young man take care of their electroplating needs.

Ralph was so successful that by the time 1923 rolled around, and he was 19, it was clear to everybody that if something drastic wasn't done, Frank and Kunigunde were never going to get their Hudson back in the garage. The timing seemed right, and so Ralph decided to move his operations across the Ohio River to downtown Cincinnati, Ohio, in the heart of commerce and industry, where he could grow his business.

Just one problem. Ralph felt sorry for his father. He may have not entirely respected the way his dad handled himself with his mother, but he loved him. Frank's boss' business wasn't going well, which meant Frank's income was suffering, and the investment troubles lingered on, and Ralph's mother was on the warpath. Ralph knew he was going to have to hire some people, anyway, why not employ his dad?

His dad could finally have a little stability, and the income would help not only his father, but his mom and his siblings. It seemed like the perfect plan, the only plan, really. And so Ralph asked his father to come work for him, and Frank accepted; only somehow, Kunigunde got it into her head that her husband was the one who owned the business, and that her eldest son would be working for *Frank*. When Ralph realized what his mother believed, he was stunned. But he kept his mouth shut. If his mom wanted to believe that, what was the harm?

And so Ralph cleared out the garage, and he and Frank moved the tanks and batteries to a building downtown, close to the Ohio River. Ralph made sure that Frank was paid a solid salary, while Ralph often took home no paycheck at all. They named the business the Stolle Steel & Iron Company.

The Stolle Steel & Iron Company was located at 236-238 Broadway, across the street from the prestigious Queen City Club, a meeting place for the rich, yet another reminder of what Ralph wasn't. Near the river, the Stolle Steel & Iron

Company was right about where, years later, a future baseball stadium would be located: Riverfront Stadium, which was later renamed Cinergy Field, which is now the west plaza of the Great American Ball Park, where the Cincinnati Reds play.

Dick Pope, who fifty years later would become president of the Stolle Corporation, suspects that where the Stolle Steel & Iron Company stood is now either somewhere under that plaza or part of the busy thoroughfare Fort Washington Way. But when Ralph moved into the Stolle Steel & Iron Company, it was in the thick of industry, and a good walk from a baseball game. (If Ralph ever managed to catch a Reds game, he would have been walking across downtown to Redland Field, at Findlay and Western Avenues, which beginning in 1934 and for years afterward was known as Crosley Field.)

The Stolle Steel & Iron Company was on the third or fourth floor of its building, according to Dick Pope, who remembers Ralph talking about those early days, when all of the electroplating equipment had to be hoisted up to the windows, because it was so heavy. They were located in a jungle of similar-sized brick buildings, and the smell of the nearby slaughterhouses was always thick with pig guts and pork loin. (By 1929, the peak of piggery in Cincinnati, almost one million hogs were being slaughtered each year.) Warehouses near Ralph's company belched smoke into the sky and dumped refuse into the Ohio River, enough that it would have been equally environmentally-toxic to have dumped 720 dead horses into the water every day. Despite the annoyance of being on a higher floor, Ralph was grateful; the river flooded several times in the 1920s, washing into the building's first floor, but several stories up, he and his workers remained dry.

Ralph was so successful that . . . it was clear to everybody that if something drastic wasn't done, Frank and Kunigunde were never going to get their Hudson back in the garage.

Ralph probably considered himself lucky that he could even get to work. Years later, he would tell his friend, Dr. Lee Beck, that he had an old car he drove across the bridge to the Stolle Steel & Iron Company every day. If the car was considered old in the 1920s, it must have been quite a relic. Beck is pretty sure it was a Ford. The car would have a blowout on almost every trip, and so Ralph became adept at fixing flats, in a matter of minutes. But there were other eccentricities about the automobile. When Ralph drove to night school, his classes were at the top of a steep hill, and his car would stall because all the gas moved to the back of the tank. Ever resourceful, Ralph drove up the hill backwards.

For a stretch of time, Ralph took the street car to get to night classes. Every evening, a woman passenger with a baby boarded the street car, and every day, she

breastfed the infant. He was startled when on one occasion, the child started crying and wouldn't pay attention to her breast. Finally the exasperated woman said, "OK, Albert, if you don't stop crying now, I'm going to give this to the conductor."

But Ralph was having trouble getting to his class, no matter what mode of transportation he used. The demands of running a business meant that Ralph couldn't keep up even his part-time pace at the University of Cincinnati, though he continued his studies at the Ohio Mechanics Institute. Years later, as it turned out, the Ohio Mechanics Institute would be swallowed up by the University of Cincinnati. It's known today as OMI College of Applied Sciences and is located in an impressive building a couple miles away from the main campus of the university.

But the Ohio Mechanics Institute has always been an impressive institution, and it certainly was impressive in Ralph's day. It came to life in 1828 as a learning center where you could pay two cents for a lecture on chemistry, physics or nature. And if you shelled out $30, you could receive lectures for a lifetime. But as the institute grew, it grew in stature, and its guest lectures included P.T. Barnum and its students included a young Thomas Edison, who studied here when he lived in Cincinnati while working for Western Union. In the aftermath of World War I, when Ralph was attending, he would have been rubbing shoulders with young men from 17 other states as well as England and Canada.

But it was a struggle to complete the two-year course. Ralph started in 1921, and it took him four years to finish the demanding program, which kept full-time students busy from sunrise to sundown. A typical day had engineering students taking mechanical drawing and descriptive geometry from 8:30 to 9:20 a.m., and algebra, plane geometry, and trigonometry in the class period following, and physics right after that—all of this before lunch. During the afternoon, students were in lab until dinner.

To receive his education, it was costing Ralph somewhere in the neighborhood of $110 a year, which paid for three classes. Underscoring the formidable equipment that Ralph was working with, when he did make it to lab, came this formidable warning from the course catalog: "Since the element of danger cannot be eliminated entirely from laboratory and shop operations, all students are urged to exercise care in the use of all apparatus and machinery and in getting on and off the elevator." Elevators, of course, were not what they are today.

The university, apparently aware of the challenges facing students like Ralph, instructed its readers: "Students are advised against trying to support themselves by outside work while attending day classes." Ralph ignored the suggestion. And in 1925, when he finally graduated from the Ohio Mechanics Institute, he was no worse off, and his business certainly hadn't been harmed. He had added rust-proofing to the firm's abilities, was employing ten workers by now, and operating with a profit. And somehow, he had also managed to take an international correspondence course in architecture.

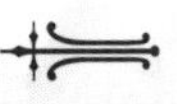

But not all was well in the business. More and more, Ralph felt his influence in the company diminished by the rest of his family, more and more of whom were joining the business. Ralph's younger brother, Howard, came aboard, as did their cousin, Paul Pendery. As far as Ralph was concerned, this was *his* business. He found it in an ad, raised the money to buy it, got it started, and he was running it. But Frank had bought into the fantasy that this was *his* business, and Ralph's mother clearly saw it that way. It disappointed Ralph that his mother didn't appreciate what he had done, but he didn't want to make his father look bad in front of his mother. So in the interest of not tearing the family fabric any more than it already had been, he left his job and, with his lifelong pal, Reuben Aschenbach, he started another enterprise altogether.

They both purchased a company in Dayton, Ohio—The Dayton Plating & Manufacturing Company, located at 26-28 South St. Clair Street, which was promptly renamed the Dayton Plating Company. Ralph was the president, and Aschenbach's title was secretary.

Twenty-one years old and an established businessman, Ralph was able to get a $5,000 loan now, and the business was a solid one. It had been around for fifty years and was a leader in gold and silver plating. Ralph was the only one in the area with a sales license for chrome plating, and so he added that to his new Dayton business, along with rust-proofing. He seemed poised to become bigger than ever.

But something happened in 1927. In an undated document titled, "History of the Stolle Corporation and the Ralph J. Stolle Corporation," the anonymous author revealed that the Cincinnati and Dayton operations were combined into one larger, more efficient plant. The headquarters was to be in Cincinnati. That's because after two years of watching the Stolle Steel & Iron Company suddenly veer into unprofitable waters under the reign of Frank, Kunigunde had an epiphany—maybe her oldest son actually had been the one making the business successful. Maybe Frank wasn't *really* running things. Maybe Ralph better come back. Now. And so she convinced Ralph to return and help the family bring the business back to life.

Maybe it was just as well. The Dayton business wasn't all that Aschenbach had hoped for. He had several mouths to feed and wasn't seeing his family much. Many nights, Aschenbach found himself sleeping on Ralph's sofa. His wife Charlotte eventually told Reuben: "You can't work with that crazy guy any more. He is going to work all of us to death. You can't keep up with him."

And so Ralph bought out Aschenbach, who in 1928 moved his family north when he took a job as a pattern maker in Sidney, Ohio, working for a company called Sidney Patterns, creating wood and metal patterns for foundries. That a spouse had helped derail Ralph's plans didn't help him come to the belief that wives

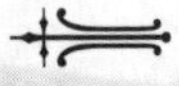

should be involved in their husbands' businesses. But while the corporate partnership was through, the friendship wasn't. Ralph would be close friends with both Reuben and Charlotte, until their deaths in the 1990s.

Right about this time, Ralph was in the midst of making a life change that would give him a life similar to Reuben's. He was courting a young woman from his Ft. Thomas neighborhood, and with making changes to his business and returning to the Cincinnati area more often, the distant sound of wedding bells could be heard. And the sound was getting louder.

The young woman was Dorothy Jane Judkins, an attractive 20-year-old brunette, with the popular short bob haircut of the 1920s. She looked young and unworldly, but she had already experienced more agony than most women twice her age. Her brothers had both died tragically. One brother, Joel, wasn't quite 20 when he succumbed to tuberculosis on February 20, 1913, at 4:10 p.m., in his family's Bellevue, Kentucky home. The younger brother Robert, born in 1895, died six years later. The cruel twist here was that Robert had actually been on the rebound from the same dreaded disease. After a fight with tuberculosis that lasted two years, which included the drastic step of moving to Phoenix for the warm climate, his health had markedly improved to the point that the tuberculosis had disappeared, and he prepared to move back home. Robert, had even written in his diary, "Since June 1, 1917, confined with TB. Now on High Road to recovery." It was not to be. He is believed to have caught swine flu, a strain from the Spanish Influenza pandemic that killed over 20 million people throughout the world in 1918 and 1919. It was a deadly disease, in which you could wake up in the morning and feel fine, be sick by noon, and dead by nightfall. Robert died in January of 1919.

FLORENCE JUDKINS HOLDING DAUGHTER, DOROTHY, RALPH'S FUTURE BRIDE.

Dorothy was a heartbroken 12-year-old about to be saddled with more grief. Four months after Robert passed away, her father was killed. Everybody, from his friends and family to his co-workers on the railroad, called him Carlos, though he was born Carolus, and on April 13, 1919, Carolus Nicholas Judkins was crushed between two trains just a few weeks before his 53rd birthday. If Carlos had lived, it

was said, he would have been able to do nothing more with his life than breathe and stare into space, comprehending nothing.

Dorothy's mother, Florence, became a seamstress, living with her daughter in an attic and struggling financially, paying rent to a family who lived in the rest of their house near the Stolles. Their bad luck would turn to some good, at least for Dorothy—she met and fell in love with the neighborhood's young go-getter, Ralph Stolle. Having a future millionaire son-in-law meant that Florence eventually wouldn't struggle with money, but she became increasingly bitter over the years by the way life had treated her family. She never remarried, eventually moved to Dayton and lived to be 94 years old.

Dorothy and Ralph had a courtship punctuated with a lot of awkwardness and uncertainty, with Dorothy doing most of the pursuing. Their old friends, George and Lois Henkle, would later reminisce that Dorothy often chuckled over Ralph's awkward manner around her. "He was so shy," she would say.

"You always got the impression that Dorothy pushed things along," says George Henkle. "And I don't know why that would be. Ralph had three sisters, so he was around girls. But I was the same way. Shy. I only became a salesman because I had to eat, and I guess Ralph was the same way. Dorothy didn't comment just once on Ralph's shyness but several times on how shy he was. I eventually put two and two together."

"She married him," says Lois, implying that they believe Dorothy might have been the one to suggest the two marry. If nothing else, Dorothy pushed the marital agenda.

The blessed day happened on October 15, 1927. As Ralph had been at Aschenbach's wedding, Reuben was the best man at Ralph's. Both mothers signed the license as witnesses, but the wedding certainly didn't happen because of Florence's encouragement. Dorothy's mother didn't want her to marry Ralph, possibly because he was taking away the last important person she had. And it didn't help that the stubborn woman was about to get an equally stubborn son-in-law. Before the wedding ever took place, Florence predicted to Dorothy: "Ralph will never amount to anything."

And shortly before the wedding, Ralph, who wasn't amounting to much, lent Florence $2,300 to pay a mortgage on a house, and get her out of the attic. But that gesture of kindness didn't make her think much of Ralph. She never much liked him, which was odd since everybody liked him, and so they never really got along.

Ralph and Dorothy's wedding was followed by a storybook honeymoon, including horseback riding, a boat trip to Lake Erie, and—it is believed—a visit to Niagara Falls. Then they moved back to Fort Thomas and, at some point, they settled into a house on 126 Highland Avenue. That they had a house, and not an apartment, to settle into was in large part thanks to Ralph's negotiating skills. A real estate agent had called Ralph and told him a house that he had admired was for sale. Ralph knew that he couldn't afford the house, and so he made an extremely low offer, and figured that would be that.

Several weeks later, the real estate agent returned to Ralph and said, "I will accept your offer." Ralph winced. Even his low offer was too much money.

"OK," said Ralph, adding, probably as a joke, "but you have to include all of the furniture." Another week or two passed, and a surprised Ralph learned that his new offer was accepted. The house was not only fully furnished, it was furnished with expensive furniture. At that point, Ralph decided he *could* afford the house.

Dorothy maintained the home, and Ralph back went to work on his business with Frank and Howard, which they incorporated on January 30, 1928, and would from then on be known as the Stolle Corporation. It didn't change much about the business, except that records would have to be better kept, and incorporation must have served as a morale booster and a signal that the company was on the right track. They also settled into a new location, a manufacturing plant at 226 W. McMicken Street. It was much farther from the river and they wouldn't have to climb all those floors. But Ralph would always had fond memories for the setting where he had first started, and many decades later would show his friend Dr. Lee Beck the place several times. Meanwhile, the Stolle headquarters at McMicken is now a children's playground.

Ralph and Dorothy began a lifestyle fairly typical for a couple in the 1920s, but it wouldn't remain that way for long.

It is not known how he caught it. Ralph himself probably never knew. Tuberculosis is spread by people already infected by the disease, coughing and sneezing their germs onto others, and so Ralph could have gotten it anywhere. What is known is that Ralph came down with tuberculosis shortly after his wedding, and he was sick for approximately a year, spanning 1928 and 1929. Dorothy must have been sick herself, aware that a disease that had hobbled her brothers now had her husband. And neither could have been comforted when Ralph's cousin, Vera, moved into Frank and Kunigunde's home to live for good. Vera's father had been killed in 1925, in what is intriguingly described as a freak accident around the house, but her mother had died in 1928 of tuberculosis. She caught the disease, and when a lesion burst, it caused her death.

Ralph was the host of a disease that had been around since before the times of the Egyptian pharaohs, a sickness that was killing off people around him, and that had been doing so effectively for at least 5,000 years.

In Ancient Greece, the famed doctor Hippocrates observed that tuberculosis was the most widespread disease of his age, and as it passed down through the ages, it has affected untold numbers of names we'll never know, and many of those we

do. Author George Orwell had tuberculosis, and Eleanor Roosevelt suffered from it. It ended the life of playwright Anton Chekhov, and earlier, Wild West gunslinger Doc Holliday, who in the last year of his life spent several months in bed, delirious for two weeks before finally dying.

One of every seven people in the world was being picked off by tuberculosis when death took a Holliday, and through the centuries, TB killed more people than the Black Death or leprosy ever did. It's estimated that in the last two centuries alone, tuberculosis ended the lives of one billion people around the world.

And now it had Ralph by the throat.

Quite literally. Some people developed tuberculosis in their muscles and joints; others had it in their lungs. Neither situation was much better than the other. To alleviate the pain in his throat, Ralph was encouraged to eat six raw eggs a day, on the orders of his doctor, who was probably Dr. Clay Crawford, the family doctor for his parents. Dr. Crawford evidently had no fears of his patients contracting salmonella, a bacteria that had been discovered in 1885 but wouldn't be associated with raw eggs for years to come.

Ralph made a deal with God: If he was given his life back, he would never again take it for granted.

Nobody who knew Ralph back then is alive to share, or can recollect, first-hand details of what he went through, but it couldn't have been pleasant, judging from the descriptions Thomas Dormandy offers in his book *The White Death: A History of Tuberculosis.* Those stricken with the disease could expect "a harsh cough, hoarseness or loss of control of the voice, an audible wheezing sound, shortness of breath on exertion, and most dramatically, haemoptysis or the coughing up of blood. Perhaps the most common manifestation was fever, usually slight but characteristically recurring in the afternoon or at night."

Dormandy added that the person with TB could generally expect to become soaked in sweat and feel physically drained yet mentally alert.

Ralph was mentally alert. Frank and Howard were running the business while Ralph lay in bed, from where he kept tabs on how their business was going. And from what he was hearing, the business was sicker than he was.

The news was disheartening. The Stolle Corporation wasn't just employing family members. Each Stolle was making $3,600 a year, and they were spending another $20,000 on their other employees, and so they couldn't afford to slack off. Ralph was young, but old enough to know that making payroll without enough productivity is any company's downfall. Meanwhile, there seemed to be no shortage of other expenses, from the coal the Stolle Corporation had to purchase for heating and powering their plant (in 1928, they spent $2,277.27) to the aluminum that they had to purchase, along with other materials (total cost in 1928:

$14,411.43) to even such matters as the legal expenses of incorporating ($150) and advertising ($754.74). They wasted money they could ill afford to, spending $955 as what were listed as bad debts. All in all, after deductions, the net income for the Stolle Corporation in 1928 was only $199. In 1929, the Stolle Corporation would have a net income of $3,723.84. That sounds like an improvement, but in reality, the company was in even worse shape. The Stolles—Ralph, Howard and Frank—didn't take all of their salary that year, a total of $16,479.03. It also didn't help matters that just as Ralph's health was improving, the economy was getting sicker. The Great Depression began October 29, 1929.

And so in the late 1920s, from Ralph's perspective—which was usually the ceiling of his bedroom—the costs of running the business were exponentially increasing, and sales weren't. Bob Mays who was born in 1928 and who would later work for Ralph explains it well: "The company was going down the tubes."

Something had to be done, if this company, which he had been nurturing for half a decade, was going to survive. And so for a long period of time—weeks, certainly, and possibly months—Ralph ignored Dr. Crawford's orders of sustained bed rest. Ralph would wake up in the early morning, drag himself to the manufacturing plant on McMicken, and throw his energy into bringing in new clients and making decisions that would keep the business surviving, if not thriving.

Then he would return home for the afternoon and collapse in bed.

Roger Brautigam, one of Stolle's future employees, came down with tuberculosis in the 1950s and was sent to a sanitarium with the prediction that he had two weeks to live. And so Brautigam has some perspective on what Ralph went through. Brautigam had the disease longer—three years—but he says that he had it better because he was given an antibiotic that had been used to treat TB since 1944. He feels he had it made compared to Stolle, who "had it at a time when there was no streptomycin, there was no wonder drug. It was strictly bed rest, very depressing. You laid there, and you knew that there was no future at all for you."

Sure enough, Ralph was in bed a lot, both at home, and in the later months, as he recovered in Drake Hospital. And it was in bed, lying flat on his back, that he made a promise. Ever the businessman, Ralph made a deal with God: If he was given his life back, he would never again take it for granted. Ralph promised that he would spend the rest of his life helping the downtrodden, in essence, giving a helping hand to those flat on their backs. It was a promise he would keep.

If he lived. Not knowing what was in store for him, Ralph lay in his bed, clinging to hope and following the doctors' instructions. He had been told that he should practice breathing as little as possible, to allow his lungs to heal. And so whenever he could, Ralph remained on his mattress, concentrating on barely breathing, undoubtedly wondering if he and his company would live to see another day, let alone another year.

1930s THE GREAT

"You must be like a monk. You must sacrifice yourself to your work."

— WALLACE STEVENS, AN AMERICAN POET WHO BECAME ACCALIMED IN 1935, BUT WHO COULD EASILY HAVE BEEN REFERRING TO ANY ENTREPRENEUR DURING THE DEPRESSION

RALPH (RIGHT) AND HIS FRIENDS AFTER A SUCCESSFUL DAY PHEASANT HUNTING AT HIS FARM

ONE MISERABLE AND WRETCHED DAY, Ralph showed up at the Newport National Bank. It was on York Street, a three-story, reddish-brick building, imposing even though a pharmacy was located in the front and the bank was squirreled away in the back.

We don't know how Ralph arrived—by car, by foot, or by the trolley tracks that rolled along York Street and past the arresting building and the ornate clock that stood watch over it from across the street. It is likely that he had sold his car by now, however. The Depression had hit Ralph hard, and during much of the 1930s, if Ralph was making any long-distance trips, he was taking a train. However he reached the Newport National Bank, this was not a pleasure visit. He didn't come to deposit money, or take any out, or even dare ask for a loan. Ralph's goals were gutted—the plans to be a prominent

businessman, all of those hopes of having a future free of money worries, the chance to show the world what he could do. He simply didn't have the resources to keep his business running.

For just as his health and that of his business had been restored, the nation's had taken a terrible turn for the worse. October 29, 1929, the stock market crashed, officially launching what would be known as the Great Depression. It's a myth that the sidewalks of Wall Street were filled with the bodies of distraught businessmen jumping from windows on that day, but the nation was plunging downward nonetheless. Cincinnati was hit just as hard as any. In 1929, 3,721 families required help meeting basic needs like food and shelter. The following year, the number was at 6,000. Two years later, 23,188 families had no dependable means of income, often living without heat, food or shelter, save perhaps the underpass of a bridge.

Campbell Graf, who years later would become Ralph's close friend, remembers his own father struggling to pay the mortgage and he himself witnessed what so many Americans have seen in old newsreels: long lines at soup kitchens and children selling apples on the streets. One of Ralph's peers and a future employee was Harold Hall, who recalls that in 1929, he was working at a railroad when the stock market crashed. He knew he was going to be laid off because of his low seniority, and so he quit and found a job in a movie theatre. He remembers entire families moving back in with the grandparents, people stuffing cardboard in their shoes after the soles wore out, and men jamming newspapers in their coats to keep them warm during the wintertime. "It was frightful," says Hall of the era.

. . . Ralph rolled the dice and gambled. He told his banker: "You can either give me an extension of my loan, or pick up the keys and run the business yourself."

And now the Depression had caught up with Ralph, who didn't even have the money to pay the mortgage on time. He rented out his house, and he and Dorothy moved in with his Uncle Al and his wife, Dot, Ralph's former classmate.

It's believed that Ralph made his journey to the bank some time in 1930. At that moment, all of his entrepreneurial dreams were dead. He laid the keys to his factory on what he believed would be its final resting place: the desk of his banker, who seemed not pleased to see him. "What in the hell are you doing giving me those keys?"

"I don't have any more money," a defeated Ralph explained. "I can't make the payroll. I can't do anything."

That included paying off a hefty loan to the bank. And instead of waiting for the bank to get cranky, or his unpaid workers to stage a mutiny, Ralph rolled the

dice and gambled. He told his banker: "You can either give me an extension of my loan, or pick up the keys and run the business yourself."

What the Newport National Bank was going to do with an electroplating company, the banker didn't know. Or maybe he had just seen too many ruined lives since the crash two years before, and he couldn't bear the thought of watching the hopes and happiness of another promising entrepreneur implode. Or maybe Ralph, at 27, seemed like a good risk in a time of uncertainty. But most certainly, the banker didn't want to run Ralph's electroplating company.

And so the banker replied, one would like to think with warmth and compassion: "I don't want the keys, Ralph. I'll loan you the money to keep your company going."

It was a good investment. Ralph kept the company going. Aware that he had just been given a chance to rise from the dead, Ralph worked harder than ever, searching for new opportunities. Meanwhile, it must have been frustrating to not see the same level of commitment from his father. In 1930, while the Stolle Corporation was in danger of imploding, its president, Frank Stolle, was running to be *reelected* to the Ft. Thomas School Board. This was a time-consuming pastime he had been pursuing for some time.

As it turned out, Frank won, on November 8, 1930. He received 1,744 votes, the most of any of his competitors. It was the same day that the hot topic of conversation in Northern Kentucky was Alma Williams, the "blond bandit," as the 24-year-old was known. Two of her friends had helped her escape from the prison hospital, and she was on the loose until November 9, when she was captured. It was the front page story in the same issue of *The Kentucky Post* that reported Frank Stolle's votes. When looking for his mention in the paper, Frank surely was aware of the blond bandit; Ralph, deep in worry over resurrecting his company from the dead, wouldn't have had a clue.

Eventually, one of Ralph's new opportunities would be in anodizing aluminum, a process that came about as an off-shoot of the rust-proofing that Ralph had perfected during the 1920s. The process of anodizing aluminum, like electroplating, requires a good degree of intelligence in the subject of chemistry. According to the web site at the Aluminum Anodizers Council, anodizing aluminum is "an electrochemical process that thickens and toughens the naturally occurring protective oxide. The resulting finish, depending on the process, is the second hardest substance known to man, second only to the diamond."

If you're scratching your head, this next bit of info probably won't help. "Anodizing converts the surface of aluminum to an oxide. While aluminum naturally forms aluminum oxide on its surface, this is a very thin film. Anodizing provides a much thicker oxide coating—several [millimeters] thick, if required," according to *Products Finishing* magazine.

In short, Ralph was a pioneer in transforming aluminum from a lightweight metal that could rust into one that couldn't, and a metal that was pliable and yet practically indestructible, inherent qualities that Ralph learned how to leverage. There was even an added dimension: it was pretty to look at.

Another of those opportunities that Ralph found, directly in the immediate aftermath of almost losing his company in 1930, came when he met Lowell Gray, a salesman for a company officially called the Aluminum Company of America, although everyone abbreviates and refers to it as Alcoa. Gray was based in Cincinnati and was soon impressed with Ralph. Somewhere around 1930, when Gray and Ralph apparently hooked up, uses for aluminum were still being discovered. Even a company like Alcoa, which had aluminum as its stock and trade, often didn't know what to do with its product, once it had been developed.

One of Ralph's earliest successes was with Frigidaire. The refrigerator company had ice cube trays in tiny compartments in their freezers, which closed with a chrome tin-plate steel door. Ralph figured out how to make these ice cube tray doors out of anodized aluminum, and to paint them, and best of all, it was half as costly as the chrome tin-plate steel. Frigidaire jumped on Ralph's idea, and was a client for years. Then, as Dick Pope says, "as the refrigerators got bigger freezers, we made those freezer doors for all of the Frigidaire refrigerators."

Ralph would simply say: "Tell me exactly what it is that you want me to do." Then Ralph would promise to do it.

Gray was impressed with successes like that, and he took Ralph up to New York City to meet Fritz Close, a salesman for Alcoa.

Close had been with the company only since 1929, his career beginning after graduating from Penn State, then spending a year working in the Oklahoma oil fields and dabbling in local Pennsylvania politics. When he met Ralph, Close was likely dressed in his trademark white shirt, dark tie, and vested suit. And the 24-year-old Close would have praised aluminum to its highest, for he often claimed that, "God—who in his wisdom had made aluminum light, abundant, lustrous, ductile, and almost boundless in its applications—was on the side of the aluminum peddler."

Close met the closest thing to his disciple in Ralph Stolle. They were just two years apart, each eager to make their marks on the world. And Ralph would simply say: "Tell me exactly what it is that you want me to do." Then Ralph would promise to do it.

What Close wanted was a tall order, literally. In 1931, the Empire State Building was nearing construction. It was a phenomenal product of its time. Great Depression, what Great Depression? It didn't destroy *all* hopes and dreams. The building was ultimately completed in one year and 45 days; workers toiled on Sundays and holidays to create a building made of 60,000 tons of steel, 2,000 cubic feet of Indiana limestone and granite, 10 million bricks and 730 tons of aluminum and stainless steel. Close was especially pleased about the aluminum, much of which was going into the Empire State Building's tower. Close and Ralph worked together on anodizing the aluminum on the tower, and their project was so successful that they next joined forces on the Rockefeller Center, most of its buildings being constructed between 1931 and 1939.

In order to fulfill his obligations to Alcoa, Ralph and his employees created the largest anodizing equipment in the world—tanks thirty feet long—in order to coat the cast aluminum spandrels, which are ornamented spaces between the right or left exterior curve of an arch and an enclosing right angle. No wonder that years later, Close would write in a letter, "Never have I had the pleasure of working with someone who had the imagination, the mechanical details, and the equipment requirements as did Ralph with his organization."

Even as the Depression emptied wallets in its wake, the Stolles didn't lose their social standing in the community, or at least they managed to occasionally maintain the appearance of living the good life. Dorothy was written up on the society page in the *Kentucky Post* in 1931, in which by this time, Ralph and Dorothy had apparently regained their residence. It was her second write-up, the first happening several months before the Great Crash. Each time, it was for hosting the New Art Literary Club. "Mrs. Ralph Stolle… entertained… most delightfully at her home Friday afternoon," stated the paper in 1929. It was equally effusive two years later, praising Mrs. Stolle for putting on a "happy affair," in which she and several other affluent women partook of a luncheon at the hostess' home, and then played an afternoon game of bridge.

Ralph and Dorothy had yet to have any children, but they would. Sandra Jane was born on September 18, 1935, and Gail Judkins two years later, on August 17. Mary Jo Cropper would arrive on May 30 in 1941. And as the Stolle family evolved, so did Ralph's company. His father, Frank, departed from the Stolle Corporation in 1933 to begin F.J. Stolle & Company with a cousin and Stolle Corporation employee, Paul Pendery. Pendery was the salesman, and Frank Stolle was the president, but at this point, Frank's main job was to sign checks throughout the next two decades. He simply didn't have the same business ambition as

Ralph, who finally possessed full control of his business. That meant steering it in some new directions, such as Sidney, Ohio. Howard didn't stick around all that much longer, though he and Ralph were always close. One by one, the family members began dropping out of the company.

The Sidney Daily News announced in March of 1934 that the Stolle Corporation would be coming to town. It was welcome news to the Depression-weary town, which was also glad to see that Joseph Kastner was purchasing the old Benjamin D. Handle factory on Broadway Avenue to provide a larger facility for his junk business, and the John Wagner Brewing Company would incorporate later that year and start manufacturing its ale once more. But if Ralph seemed like a big-wig moving into town, it was only an illusion. He was running on financial fumes. The Stolle Corporation 1934 tax form showed a total income of over $42,000, but a deficit of approximately $1,200.

But he was bringing his business to Sidney because things were looking up. His old pal, Reuben Aschenbach, had been doing some work for the Prima Washing Machine Company in Sidney, making wood patterns for castings, and he learned that they were thinking about chrome plating their tubs. Aschenbach sent word to Ralph, who hopped on the train and quickly met with Robert James Anderson, the president of the Prima Washing Machine Company.

Although it wasn't as large as Maytag or Whirlpool, Prima did respectable business and had a lot of important clients on the East Coast, and several years earlier it had advanced the laundry industry considerably. In 1927, Anderson had invented the never-crush ringer, which goes around the washing machine bowl. Before 1927, if for any reason you stuck your hand in the ring, chances were you'd never see your hand again, at least not in the way you remembered it. But Anderson's invention meant that if your hand went into the machine, it automatically stopped. He sold the patent to Maytag, which still uses the technology today, and at the time, it was said to be the largest sale of a patent yet: $177,000. Anderson received it all in cash, and it carried the company through the worst years of the Depression.

Ralph had struck up a deal with Anderson, in which the Stolle Corporation would anodize the aluminum for Prima's agitators and rings. Part of the goal was to make the metal shine, for which there is actually a term: *bright work*. It was good steady employment, and Ralph moved part of his company up to Sidney, renting out some space in a building, very close to the Prima Washing Machine Company. Literally, you could walk from Prima, across Park Street, to a building, and behind that building, on Oak Street, was the Stolle Corporation. The space wasn't big and took up only one floor, recalls Tom A. Anderson, the son of Prima's president.

The spring of 1934 gave way to the summer as Ralph put his operations together. In the contract with Prima Washing Machine Company (which officially

was the Prima Manufacturing Company, but nobody ever called it that), Ralph agreed to have his business up and running by April 15, 1934. In the contract between the Stolle Corporation and Prima, it was agreed that Ralph could solicit other business, but Prima would always receive preference. If Ralph had two deadlines to meet, Prima's always came first. In exchange, for the next four years, Prima would give the Stolle Corporation all of its plating and rust-proofing work, to cooperate as though the two companies were one, and most important, once the invoice was delivered, to pay on or before the following Friday. And in exchange for paying on time, Prima would receive a three-percent discount.

The contract was all very formal, with references to "the first party" and "the second party" and words like "herein" and "hereinafter," but it meant one very beautiful thing to Ralph: stability.

The business clearly needed it. Robert Anderson's son, Tom, remembers hearing about the year of 1934 as he grew up. "This is strictly secondhand," cautions Anderson, "but I can remember fellows telling me that when he first started here, they'd polish up an agitator or ringer base, and then they'd take it over [to Prima], and get their money, and then come back and do another one." Anderson is also fairly certain that his father co-signed a $2,500 loan, to help get Ralph up and running in Sidney.

Ralph and his employees worked hard, and it couldn't have helped matters that a punishing heat had settled over Sidney, Ohio. Rain was in short supply, and temperatures were in the mid-90s in May, which would seem spring-like once the mercury passed 100 throughout July, July, and August. Farmers watched their crops die. The local scene wasn't as bad as those in Texas and Oklahoma, where the Dust Bowl gripped the farmers and their families as well as the nation's imagination, but in Sidney, brows were sweating and the backs of shirts were sticky. It was the type of summer where you never knew quite what would happen. In the middle of August, a woman's body was found bunched up in the weeds of the Great Miami River. The coroner ruled it a suicide. That same week, a fight broke outside of a dance club north of the city, and Harley Fry was the recipient of a fatal blow. Meanwhile, an elderly man, Sol Springer, was found alive near Tawawa Lake, missing for four days, having wandered away from home.

In the midst of such drama, Ralph was spending days at a time in Sidney, sleeping in a room at the Aschenbachs' home, then taking the train back to Cincinnati to be with Dorothy during the weekends. Fortunately, as the Sidney operations started to stabilize, he could occasionally afford to not always be in Sidney during the work week. Jake Zwiesler, Ralph knew, would handle any problems. Zwiesler handled everything for quite awhile, including running all of the company's errands that required hauling heavy loads, because Ralph couldn't afford a delivery truck. He sped aluminum and other material back and forth in his Ford automobile. Zwiesler was one of a handful of men recruited by Ralph to

work in Sidney. He had put an ad in the *Dayton Daily News*, and when Zwiesler telephoned him, Ralph's first words were: "I was sure hoping you'd answer that ad." Zwiesler had worked with Ralph at the Cincinnati electroplating company in the 1920s, and Ralph knew he was just the person to supervise operations in Sidney.

Prima was a big part of Ralph's comeback. The Empire State Building and Rockefeller Center had been exciting but temporary work. This was ongoing, something that could help sustain the employees in Sidney and truly kick back substantial income into the Cincinnati operation. After a dozen years in business, the Stolle Corporation had finally achieved stability, and in the midst of the Great Depression no less.

But the same couldn't be said for Prima. In 1935, Prima went bankrupt.

Almost as quickly as it had happened, the Stolle Corporation was left without a major source of income, and because the contract had obliged them to make Prima their best customer, that's exactly what they were; without the washing machine company, the Stolle Corporation was left with a gaping hole in its earning potential. Ralph rarely, if ever, showed his friends or family anything less than optimism, but one would have to suspect that he had a few moments late at night where he wondered if his company would follow Prima into the grave.

Robert Anderson never really bounced back—by 1937, he had lost his home and just about everything else; Ralph, however, was able to recover pretty quickly. But only by scrambling for dear life, and because of a chance meeting on a train. Some time after Prima's financial washout, Ralph was riding back to Cincinnati when he began talking to the fellow next to him, who happened to be a purchasing agent at Frigidaire. The two men got to talking, and before the hour and fifteen minute trip was over, it was pretty clear that Ralph had found his next big break. The purchasing agent told Ralph that they were looking for a way to add space to their operations, but without adding any new buildings. They needed to outsource. Ralph didn't waste time, telling the man: "I will do the work for ten percent cheaper than you are doing the work now."

> *"Ralph would tell us to not use the word, 'I.' Everyone fell in behind that. We all used the word 'we.' "*

"Well, you come on down, and it's yours," the agent told Ralph.

It was a big break, and Frigidaire sent a lot of work to the Stolle Corporation. Meanwhile, Ralph searched for new uses for anodized aluminum and realized he could replace the nickel and chrome on automobiles and appliances.

It wasn't that simple, though. As bright as aluminum was, it wasn't bright enough. Ralph needed aluminum that was at least as attractive as nickel and chrome, if not more so. And so Ralph—or possibly one of his engineers, though

it's an idea that's always been credited to Ralph—came up with a process that did just that. Quite possibly, it was a process that both Ralph and his workers came up with, since credit tended to be shared among the staff. As Pope observes, "Ralph would tell us to not use the word, 'I.' Everyone fell in behind that. We all used the word 'we.'"

In any case, the process worked so well that Frigidaire converted all of its refrigerator trim to bright aluminum, presumably putting out of business untold numbers of nickel and chrome entrepreneurs, and the Sidney factory had to be transformed from a plating and finishing shop into a full-throttle manufacturing plant. Meanwhile, Prima was purchased by a new company, one that also needed agitators and rings, and Ralph's business was able to furnish them as well.

And never again would Ralph's company depend so much on one stream of revenue. In fact, that was one of the unique things about Ralph's company, observes Pope. "It had to reinvent itself almost every year or two. That's because the parts that they made for cars or appliances would change. They would redesign the refrigerator or store. In Ralph's heyday, they were redesigning cars every year. They would change the whole bodies. So the grills and other components of the car were changing rapidly. So when you went out to bid on a job, you were only going to have for a year or maybe two. Then you had to go back and bid on other jobs in order to keep the plant running. It was a constant battle all the time."

One April Sunday in 1939, Ralph was given a reminder that his new plan—constantly evolving and looking for new opportunities—was a sound one. If Ralph had decided to stick with washing machines and only washing machines, this would have stomped him out for good. The building that once housed the proud Prima Washing Machine Company caught on fire. Tom Anderson was 12 years old and on his bicycle when he saw the cloud of smoke rising into the sky. He pedaled in a hurry to see what was burning and was surprised, and not cynical or mature enough, to be particularly delighted or saddened to see the building go. He was just fascinated. "It was a big one," says Anderson, who recalls that one of two adjacent buildings had a lot of paint thinner in it, and the fire department was especially alarmed that if the flames reached the thinner, the building might detonate.

The situation was no less tense across the street where Jake Zwiesler was handling everything, just as Ralph knew he would. Zwiesler was on the roof of the building. Norma Counts, his daughter, remembers the scene vividly. Her father was with another colleague, and the two men were hosing down the roof. She had been with her father at home when he received a call that the Stolle plant was in flames, and from that moment on, everything moved at top speed. "Dad and him were watering it down trying to save Stolle's," says Counts, who was then 12 years old and in a growing crowd of people who had come to watch, all of them possibly unaware of the impending danger. "The building had a lot of chemicals inside," says Counts. "If the fire had reached it, there would have been quite an explosion."

The Prima building burned to the ground. The Stolle Corporation was untouched. One of the lessons that Ralph had long since learned—surround yourself with good people—was paying off.

As the year of 1939 and the worst economic decade in American history slogged towards a close, Ralph had much to be pleased about, and, indeed, he felt comfortable enough to start relaxing a little and taking an occasional evening constitutional in Sidney. Maxine Pence remembers seeing her boss in the town square in Sidney on occasional Saturday nights. She had just started her job as an inspector, and she would see Ralph walking around, greeting employees. He was still working, really—being seen and seeing his employees was an important human resources touch—but it was a more laid back affair.

Everybody who was anybody was generally in the town square during the evenings, says Pence, and while there were numerous people from various companies, the Stolle Corporation was well represented: The company was, points out Pence, one of the biggest employers in Sidney at the time.

And if you were living and working in Sidney then, you parked on the edge of the square, walked around, socialized with friends and neighbors and then did your grocery shopping. And sometimes Ralph would be in there, stopping to socialize himself.

"He knew everybody who worked for him, and he knew the kids," says Pence, who left the company in 1942 to raise her three boys but came back in 1958 as a secretary and retired in 1983. But it's those pleasant memories in 1939 that have really stayed with her.

"Ralph made you feel important," says Pence. "If your family was sick, he let you off. Family was always first for him. And he knew that if your family was in good shape, the workers were in good shape." And, indeed, Ralph always took care of his workers. He always seemed to be trying to take care of everybody, especially if they had taken care of him. Robert Anderson is a good example. He didn't necessarily need Ralph's help; he picked himself up after the hammering that the Depression gave him, though he would never reach the heights he once climbed. He would do some consulting work, and he would make a reasonable living for himself, his wife and their five children.

But if he had struggled, Ralph would have been there. He never forgot what the president of Prima had once tried to do for him. As far as he was concerned, Prima would forever be part of the Stolle family. Years later, Tom was told by the Stolle Corporation's personnel director: "Ralph always said, 'Take care of the Anderson boys, if they need a job.'"

Chapter Four

1940s RALPH'S

"You can't create an entrepreneur. There has to be an internal, inbred spark. Once you recognize that, you're in business. And Ralph had it, and recognized it."

— LONG-TIME COLLEAGUE AND FRIEND BOB MAYS

RALPH LOVED HIS COWS. HERE'S ONE OF HIS PRIZE HOLSTEIN HEIFERS.

IF YOU KNEW RALPH, YOU KNEW OF HIS FARM.

The site in Lebanon, Ohio, was part of the businessman's identity. Randolph William Hearst had his castle. Franklin Roosevelt had his Shangri-La (which was later renamed Camp David by President Eisenhower). Thomas Edison, Menlo Park. No rolling of the eyes, please. It's just an analogy, and it is doubtful Ralph ever considered his farm comparable to such legendary real estate. He wore modesty like an old pair of slippers. Still, locally, he would become something of a larger than life figure, and certainly his farm became his castle, his retreat, and his laboratory. There was no farm at first, or a house to live in, just land—lots of it—stretching over acres and acres. Eventually, the farm would be 3,500 acres of rolling hills and meadows and a lovely little lake. It also fulfilled one of Ralph's childhood ambitions. He was a farmer.

The property was technically in Oregonia, Ohio, but it had a Lebanon address, and Ralph called it San-Mar-Gale farm, named

after his three daughters, Sandra Jane, Mary Jo, and Gail Judkins. He bought it on May 30, 1941, the day his third daughter, Mary Jo, was born. That he apparently wasn't in the waiting room was a combination of his workaholic nature and just a sign of the times: Giving birth was for women; and for men, making sure the family had a place to live. That said, they wouldn't actually move to the farm for another twelve years to the day—on Mary Jo's birthday. In between the day of her birth and her 12th birthday, Mary Jo recalls, they spent every summer at the farm, and many weekends. "We even ate dinner there some week nights, and then drove home."

Mary Jo marvels at that, because these were the days before the Interstate, and the drive back to Fort Thomas, she surmises, was over an hour. "But that's how Dad was," she reasons. "He always had to be doing something."

Indeed. Frequently, Mary Jo, Gail, and Sandra wouldn't see their father for days at a time. He arrived home after they had gone to sleep and was out of the house before they awakened. Sometimes, he didn't come home at all, opting to spend a few nights in a hotel. Work kept him busy, and yet he was still a presence at the house. Mary Jo remembers the family gathering around the TV to watch Milton Berle—and her father looming nearby, perhaps not quite as interested, but there. Ralph, in fact, rarely appeared interested in television, although in 1955 *The Lawrence Welk Show* began airing, and that was a show he eventually took to, and he always enjoyed boxing, a sport frequently airing during the early days of TV, the late 1940s and early '50s, in part because of the technology factor: the confined space of the ring made it easy for the cumbersome cameras to focus on the action. "Daddy loved to watch boxing," remembers Mary Jo. "He would get so excited, he would almost do the fight, throwing punches along with the boxers."

Eventually, the farm would be 3,500 acres of rolling hills and meadows and a lovely little lake. It also fulfilled one of Ralph's childhood ambitions. He was a farmer.

But generally, Ralph wasn't interested in television at all. He rarely went to the movies. He did go to a few Broadway plays in New York City on some family weekends, and so his fleeting interest in drama during high school hadn't necessarily completely died out. But that was a rare foray into the arts and popular culture. "He was moving too fast to do much of that," explains his old friend, George Henkle. "You don't have time to write letters or read books, when you're on that circuit."

But he wasn't completely ignorant of popular culture. Henkle says that Ralph kept abreast on the changes going on around him—*I Love Lucy*, Elvis, the

hula hoop—because his daughters kept him informed. And Ralph occasionally would even let himself fall into a fad. At least once. Although he was always impeccably dressed, his wardrobe took a little turn for the casual during the hot months of July and August, recalls Rita Hauck, who went to work in Cincinnati for the Stolle Corporation in 1953 and remembers seeing Ralph periodically appear in the office.

During a time before air-conditioning was completely pervasive, Ralph would turn up in the office in a seersucker suit with little blue and white stripes. "It was the 'in' thing back then, although he didn't look quite right in it," admits Hauck.

It could be that Ralph simply wore the seersucker suits because the light cotton cloth was comfortable in the humid Ohio summers when air-conditioning was popular but not yet routine. But Ralph had surely seen these suits being worn around the campus during his brief, abridged college years; university men were just starting to wear the seersucker shortly after World War II, right about the time Ralph was grabbing day time courses and taking every night course he could find at the University of Cincinnati. Ralph wasn't rich then. If he couldn't afford to wear a seersucker suit back then, he certainly could now.

Besides, the fabric had picked up popularity in posh country clubs in the humid South during the Great Depression, and after World War II, businessmen up north were discovering the new status symbol. The suit symbolized business, which suited Ralph just fine.

As his granddaughter, Cathy Chasteen, suggests, "I think Grandpa threw himself into whatever role he was expected to play, whether it was grandfather, father, or whatever." But the role of businessman, Ralph played to the hilt. He was a businessman who happened to be a grandfather, father, husband, or friend. He was so locked into his business persona that he apparently didn't own a pair of short pants, not even when he went vacationing in Florida. "I never saw him in shorts," says George. "I think he dressed up a bit on all occasions, even for close friends. I can't remember him in shorts, even when it was hot. But then he was from the 'old school.' My father didn't wear shorts."

But if Ralph was all business during the 1940s and early 1950s, he still managed to keep the homestead interesting. He enjoyed telling tall tales. For awhile, he had his daughters believing that he ran track with Jesse Owens, until they started doing the arithmetic and realized that, despite the fact that geographically Owens was from Ohio, there was no way the two would have been competing against each other. Ralph also relished telling how he and Dorothy had been driving in the Kentucky hills during their courtship in the 1920s, when they wound up getting between feuding bootleggers. Bullets were spraying everywhere, and Ralph's ear was nipped by one of the rounds. While the daughters doubt that, this story actually was basically true, according to their mother, who wasn't one for spinning tales.

"Daddy would bring a calf or rabbits home from the farm. There were always

some animals around the house," recalls Mary Jo. "Once, he brought two horses to the house, but they escaped, and so we had two horses roaming the streets of Fort Thomas. And when the police found them, they knew exactly who to call. We must have seemed like the strangest family."

But Mary Jo never fully realized just how odd or fortunate their family must have appeared to outsiders. She didn't understand that they were rich. When her father learned that his buffalo were roaming about on their Lebanon farm during a blizzard, she half-assumed—if she thought about it at all—that her friends' parents were encountering similar problems. But as the years went on, Mary Jo started to understand that not every child had a father like Ralph. Upon graduating from high school, Mary Jo was given a two-door, all white Ford Fairlane, which she adored. But six weeks later, she backed into a pole at her uncle's home, damaging the fender. Ralph was not pleased. But he said he would pay to have it fixed if Mary Jo would let him experiment with some aluminum panels that he thought might work on cars.

"That was a no brainer, until I went to pick up the car from the dealership where it was being repaired," recalls Mary Jo. "The panels were an awful turquoise color, and looked like big tail fins on both sides." When she went to college at Ohio State University, and whenever she drove around Lebanon, people recognized Mary Jo and her mode of transportation. "I could hardly wait to get rid of that car!"

But Mary Jo never fully realized just how odd or fortunate their family must have appeared to outsiders. She didn't understand that they were rich.

When Ralph wasn't paneling his daughter's car, raising his buffalo, or interrelating with the family in some other way, he was working. And when he wasn't working, he was driving to work. Sometimes, he traveled on long business trips and brought the girls along. Mary Jo has vivid memories of making paper dolls with her two sisters in the back seat of their car in the early 1940s, while her father would be inside the Frigidaire or Whirlpool headquarters, sealing an important deal. But mostly, Ralph was shuttling back and forth from the Stolle Corporation in Sidney and Cincinnati and to his web of family roots in Fort Thomas, from Dorothy and the kids to his parents and many of his siblings. And so for a long time, he kept commuting to Sidney, a distance of approximately a hundred miles.

Today, it takes about an hour and 40 minutes to drive from the two cities, but that's utilizing I-75 and I-275. During the 1930s and 1940s, there was no freeway system, and there wouldn't be until President Eisenhower came along. So to get where he needed to be, Ralph took the train, and later drove along well-traveled countryside roads. Much later, recalls his old pal George Henkle, Ralph would tell his friends about those days.

"But when did you sleep?" somebody asked.

"Oh, I slept at the stop lights," said Ralph. "When it turned red, I dozed."

"But if you were sleeping at the lights, how did you know when to move on?" somebody else asked.

"Oh, they'd always blow the horn at you," explained Ralph.

He wasn't kidding. When he brought his daughters on trips, he'd tell them to wake him up when the traffic signal indicated it was time to go. Ralph loved power napping, and on longer trips, in particular visiting the Alcoa headquarters in Pennsylvania, he frequently invited along young Bob Mays, a teenage boy who lived in the neighborhood. Ralph liked the lad, they became great pals, and eventually Mays worked for Ralph. But Ralph also wanted Mays around for the company, somebody who could keep him awake.

On one occasion, Ralph was jolted awake by the blaring of automobiles honking, and the businessman glanced back to see what looked to be a half-mile of cars behind him. Soon afterwards, he bought an airplane.

As Ralph settled into the routine of successful businessman, his father, Frank, was getting on in years. He had been slowing down throughout much of the 1940s, and in 1949, Frank retired from the F. J. Stolle & Company. That meant Kunigunde had him on a shorter leash now, but nobody remembers him complaining. Ron Pendery remembers spending time at his uncle and aunt's second home in Indian Lake, and she would send Frank out to the meat market or grocery store, and invariably he would bring home the wrong items. "And so she would make him go back and get whatever it was on the list," says Ron. "She had him on the run, pretty good." And when he had had enough, back at the house in Fort Thomas, he escaped, when he could, to his workbench in the basement.

"He was a sweet, not stuffy man," recalls his granddaughter, Mary Jo, "but his appearance was staunch. He was this little tiny fastidious person. Very proper. And I can remember the grandkids, especially Ron and Ken, playing jokes on him. They would hide in a car and whistle at girls, who would glare at Frank. They teased him a lot."

And then they mourned him. On October 16, 1952, Frank died, at the age of 74. After the 10 a.m. services at the Dobbling Funeral Home in Fort Thomas, everybody gathered to mourn the loss of the family patriarch, but there was little doubt as to whom everybody would be turning for guidance now. The family members gathered at Ralph and Dorothy's house in Fort Thomas on Carolina Avenue.

Kunigunde retired to Indian Lake. Today, it's still a popular destination but for a time in the 20th century it was known as "the Midwest's Million Dollar Playground." Over one hundred years earlier, Indians and settlers fought here, and the famous frontiersmen Daniel Boone and Simon Kenton were said to have traveled through the area. But by the time the Stolles discovered it, it was a year-round recreational paradise. While Indian Lake was a far drive from the family in Fort Thomas, Kunigunde would have plenty of visitors willing to come up and spend time with her. And there was another bonus, too: It was less than an hour's drive from Sidney.

With his father having passed on, and his mother adjusting to her new home, Ralph felt it was finally time to make a change in his own life. He called the family together and asked them if they would finally be willing to move to the farm in Lebanon. It was an idea he had floated before, but because he tended to run his family more like a democracy than a company, his daughters had been truthful: They were not enthusiastic about the plan, which would mean leaving neighborhood friends behind. This time, he presented his mission with more urgency. His commute would be cut appreciably, he explained, and he could spend more time with the family.

Sandra had already moved out of the house, and Gail was getting ready to leave for college, and so it was really up to 11-year-old Mary Jo, her mother, and her father. And this time, Ralph won out. They decided to move to Lebanon, cutting his driving commute in half, which begs the question, of course: Why not just move to Sidney?

Over fifty years later, Mary Jo has a ready answer: "Because my dad never believed you should work where you live."

Her explanation is not that surprising, considering the lessons Ralph had learned as a child. Bring your family into your business, and your business is no longer your own business. And Ralph, who could be as old-fashioned as they come, meant it. Had there been a Take Our Daughters to Work Day in his time, he would have passed.

"My father kept his business and personal life very separate," continues Mary Jo. "I remember when I was filling out my college application, I had to ask my mother: 'What should I put down as my daddy's occupation?'"

Ralph and his family easily adjusted to life on the farm. Their ranch house had been built several years earlier, from blueprints Ralph had designed himself, and, of course, the Stolles had visited here often and were intimately familiar with its surroundings. Ralph recruited numerous workers from Sidney to help with the building, and several workers sometimes slept on the site.

The house was on a hill, overlooking a lake and a dairy barn. Farther up the hill would be, in time, a small airport. It was the one thing that Ralph had—an airplane, a hangar, runway—that suddenly let his teenage daughter, Mary Jo, know that she was different from her other friends. None of *them* had a father going to work in an airplane.

It was embarrassing for the teenager, but necessary for her middle-aged father, who was tiring of losing so much time on the road. The commute to Sidney was still significant—at least an hour's drive—but with the airplane and a hired pilot at the wheel, he could make a one-way trip in about twenty minutes. And suddenly Ralph was able to be home almost every evening to have dinner with Dorothy and Mary Jo—provided one of them would turn on the running lights, so the pilot could land. And with his new toy, he could take up friends and family to show them what life was like above the clouds. Ralph was a showman, and any time he could make the world a little more exciting for somebody, he would.

FAMILY PORTRAIT. TOP BACK ROW, FROM LEFT TO RIGHT: LES STOLLE, MARGARET AND DON HILLIKER, IRMA AND ROY PENDERY, CHARLOTTE AND PAUL PENDERY. SITTING, LEFT TO RIGHT: EDITH AND HOWARD STOLLE, KUNIGUNDE STOLLE, RALPH AND DOROTHY STOLLE.

But Dorothy was an exception. As far as anybody recalls, she only flew once in her life, and that was enough.

Dorothy was much happier with her feet planted firmly on terra firma, and Ralph did his best to make sure they quickly were ingratiated into the social scene. They joined a local Lebanon dinner club, whose members were attorneys, doctors, executives, and other professionals, and they soon formed a bond with real estate entrepreneur George Henkle, and his wife Lois, which endured for the rest of their lives. “We're at this party, and here's this new couple,” recalled George some fifty years later, “and Ralph was just so interesting to talk to. His company was interesting, and his hobby was raising cattle. Everybody was fascinated with what he had to say.”

But at that dinner party, which was held at an attorney's house and entailed games of bridge and a smorgasbord of covered dishes, Ralph wasn't holding court in large crowds, but in smaller, intimate conversations. While Ralph was a showman, his personality was never anything similar to a gregarious emcee. He loved parties, but not being the life of one. He could talk for hours with a few people, but he could not talk to a large number of people for even a few minutes. And so he possessed at once polar opposites, both stagecraft and stagefright.

George and Lois began seeing Ralph's theatrics when they attended parties that Ralph held at the farm. There had been sizeable gatherings at the house in Fort Thomas, but a social event at the Stolles' farm was akin to visiting a small country. You met all kinds of people, and literally, you would want to pack your bags. George recalls that Ralph often flew in on his personal airplane some bigwigs from Alcoa or General Motors to attend one of his weekend parties, where they often hunted pheasants and quail on his game preserve. They dined on a feast of roast beef, not to mention roast chicken and roast turkey. And then many of the men finished off the night with a marathon poker game.

"They loved those weekend parties," says George. "Most executives' business trips were impersonal, bland affairs. Where else could they go for a weekend at no cost and hunt on a private game preserve? Nobody else did that for them."

The farm was, in many ways, a theme park for adults—and kids. In the 1940s, whenever Ralph's daughters would beg to be allowed to go to summer camp, their father immediately shot those ideas down. "You have camp here," he said, of their farm. But he wasn't being insensitive. Ralph was frightened of polio. The deadly disease had first invaded the United States in 1916, and by the 1930s, it was a summer ritual of the most deadly sort. For the next two decades—into the mid- 1950s,—two million Americans would be infected by the virus, which attached itself to the neurons in the spinal cord and brain stem, causing headaches, neck pain, fever, muscle soreness, and weakness, and occasionally muscle paralysis and difficulty breathing. It was believed that swimming pools teemed with the bacteria during the hottest months of the summer, and like many parents, Ralph didn't want to chance his daughters coming in contact with other children. As a tuberculosis survivor, the thought of one of his daughters catching a deadly disease must have chilled him to the bone.

[Ralph] could talk for hours with a few people, but he could not talk to a large number of people for even a few minutes.

But he had other worries. In the years after moving to the farm, Ralph began to fear for his family's safety. A doctor in the area had received a grim threat, promising his daughter would be kidnapped if a certain amount of money wasn't left on

RALPH'S WIFE, DOROTHY, WAS SHY, BUT SHE NEVERTHELESS POSED FOR THIS FORMAL PORTRAIT IN 1975.

a hillside. Ralph had plenty of money. There was reason to believe a nutjob attracted to a wealthy country doctor might as well be attracted to a millionaire businessman.

For a time, Mary Jo wouldn't take the bus to school; Ralph had Dorothy drive her there. And besides, there were the union workers ominously hovering in the shadows. They likely had no plans to harm one of the Stolle girls, but Ralph imagined some might want to, if it meant breaking their old man.

The unions had plagued Ralph as early as the 1930s—in 1937, some workers unsuccessfully tried to organize a strike—but it was the decade after World War II when things started to get particularly ugly. Ralph did everything he could to offer his workers fair wages and incentives. Everybody had health insurance, and a $10,000 life insurance plan after being at the company for three years, which was a liberal amount for the 1950s. There was even a generous pension plan; the Stolle Corporation had been one of the first businesses in Ohio to apply for one during the 1930s. Workers since at least the 1930s had been allowed a week's paid vaca-tion. As far as Ralph was concerned, he had done nothing to incur the wrath of the unions.

Throughout the 1950s, Ralph's idyllic life at the farm and his work routine were often pummeled by the unions. In June of 1951, seventy production employees left his metal plating plant at 227 West McMicken Avenue. The following month, a 33-year-old employee of Ralph's, and somebody who detested unions, was arrested for bringing a gun to a picket line, pointing it at a picketer, and then getting into a fight. The trouble with the unions frequently commanded Ralph's attention during the 1950s. George Henkle recalls a trip to Alaska that was —sort of—called off at the last minute. The Alaskan Highway had been built just a decade before, a nine-month $135-million dollar project in 1942, hurriedly created when it was feared Alaska might need a military supply route if Japan attacked the state. In 1948, the highway opened to the public, and so during the following decade, George, Ralph, and another pal, Hewett Mulford, who had a flower importing business, found themselves talking about it. George Henkle remembers what happened next:

"The Alaskan Highway had been built, and we got to talking one time about how much fun it would be to go up to Alaska and fish. We planned the trip, and Ralph bought a new station wagon, just for the trip, and we had our camping gear packed and everything, and at the last minute, Ralph said he couldn't go and would have to pull out. Labor problems. Well, Ralph told us to go on the trip without him—and to take his new station wagon. He even hired a college boy to drive us. We flew to the southern part of the highway, and the college boy met us there, and drove us to Juneau. And every once in awhile, we'd call Ralph and tell him how things were going. He wanted us to call him every day, to say how it was going, but telephones were often in short supply up there. So Hewitt and I wandered around Alaska, in Ralph's car. We did buy him new tires when we got back. Some of the highway was gravel, and, well, he needed them."

But the union troubles that had plagued Ralph for parts of a decade really got unpleasant in 1956. Despite the fact that the Stolle Corporation had nothing to do with mining, the Mine Workers Local 50 was lurking in the corridors, trying to convince the employees to vote to join their union. Ralph would have none of it, and he had a point in insisting that his employees not vote: The Mine Workers Local 50 wasn't recognized as a union under federal law. The Mine Workers lost their credibility during the most intense years of McCarthyism, 1950 to 1954, when Senator Joseph McCarthy led the charge to rid the country of communism; the Mine Workers leaders were required to sign papers stating that they had no Communist affiliation, and they wouldn't do it. And so as far as the Stolle Corporation was concerned, there would be no elections for employees to vote themselves into a union not recognized by the federal government.

Ralph's logic could be considered right or wrong, depending how one looks at it, but it wasn't shaky.

As far as Ralph was concerned, he had done nothing to incur the wrath of the unions.

Some employees thought so, however, and a crowd of them began picketing s the year of 1956 wore on. And then it got ugly.

Cars were overturned, and at night union organizers or strikers shot out plant and office windows. A mini-riot erupted, injuring two people. Ralph hired Pinkerton guards to stand on the roof wielding guns, and he sent company trucks packed with employees to drive through the picket lines—slowly, however, and not mowing anybody down. Meanwhile, Common Pleas Court Judge Beery issued a restraining order limiting the number of pickets. And then in October of 1956, somebody threw a stick of dynamite at Plant One on Oak Street.

It was night when that happened, and only a few windows were blown to bits, along with some concrete blocks, and then some more damage came with another explosion at the Sidney Truck and Storage Company location on East Poplar Street, which had supplied Ralph's company trucks. Nobody was injured there, either, but the blasts did what they were supposed to. Employees were scared, not to mention townspeople: the nighttime explosions awakened thousands of residents.

Something had to be done, and right away, Ralph had sheets of paper printed out and placed around Sidney. The paper posters read:

$5,000 Reward
For any information leading to the arrest and conviction of any person or persons responsible for the dynamiting of the Sidney Truck & Storage Co. and the Stolle Corporation Plant.
The Stolle Corporation
By Ralph J. Stolle

Meanwhile, the police hit back hard. The dynamite-wielding employees were never caught, but three employees were indicted for inciting to riot, and convicted. And nineteen more people were indicted for a "pattern of violence." Other employees were charged with disturbing the police. But it got even more ugly. The attorney representing many of the defendants, Forrest Blankenship, from Troy, Ohio, commented in the Troy newspaper that he had evidence that "Ralph Stolle himself is responsible for the violence at the plant."

Incensed, Ralph unleashed a $1 million libel and slander suit against Blankenship and the union. Soon after, an enormous bomb—consisting of ten sticks of dynamite—was tossed next to the Stolle plan on Park Street. But nothing happened. The bomb just sat there, rattling the city until an army demolition expert could be brought in to diffuse and haul it away.

By year's end, four more picketers were tossed into the clink, and as the months grew colder and 1957 arrived, fewer and fewer workers were striking. Those who did huddled next to barrels with fires lit inside. Dick Pope remembers that it all ended just before Easter weekend in 1957. There was one striking worker, marching up and down. Suddenly, he just stopped, walked over to a barrel, tossed his sign into the flames, then returned to work.

Pope says that after the unrest and hard feelings died down, Ralph took a long hard look at the houses across Plant One on Oak Street. It had been at one of these houses, authorities suspected, that many of the gunshots had come from. And so Ralph bought all of the houses across the street for $35,000 so it could never happen again. Then he tore the houses down and sold the land for $90,000. "It was a typical Ralph move," chuckles Pope.

It was the stress of dealing with the unions that ultimately drove Ralph to make several shrewd business decisions. First, he sold forty-nine percent of his stock in the Stolle Corporation to Alcoa, realizing millions of dollars on the sale.

Ralph didn't go around sharing what he had made, but it was enough to start the Ralph J. Stolle Company in 1954. Ralph remained the CEO of the Ralph J. Stolle Corporation, while also acting as CEO of the Ralph J. Stolle Company. He kept the Stolle Corporation intact, but the Ralph J. Stolle Company would be a business that owned lots of little companies—companies that would never have more than 200 employees. Because of that, said Ralph's friend, George Henkle, Stolle's companies would be too small to allow unions. Dick Pope, who ultimately became president of the corporation in the 1970s, says that the companies were designed to be small not to be union-proofed, but because they could operate quicker and faster than a bloated bureaucracy. In truth, both Henkle and Pope are likely correct.

There is no question that from a historical perspective, of course, unions have made life better for employees. For instance, in 1903, the year before Ralph was born, Mother Jones led a march from Pennsylvania to New York City to demand Theodore Roosevelt improve working conditions for children (she never met with him, but the changes were made). And in 1908, as a septuagenarian, she helped organize a strike in Paint Creek, West Virginia; during the strike, the mine-owners hired men to machine gun the strikers and their families. There had been a lot of bad blood, much of it spilled between union workers and the companies that hired them, and Ralph wanted nothing to with the strife—but, says Dick Pope, he did have a keen appreciation for what union leaders like Mother Jones had done for the country. And the mounting evidence suggests that Ralph went out of his way to make his companies so enjoyable to work in that nobody would want to join a union.

. . . Ralph went out of his way to make his companies so enjoyable to work in that nobody would want to join a union.

Meanwhile, Ralph also did everything he could to avoid hiring anybody who had ever worked for a union, or expressed any desire to ever be in one. He began recruiting employees in Kentucky and Tennessee, in rural towns that had not been enjoying the booming post World War II economy that the rest of the country was seeing. In that sense, Ralph was single-handedly responsible for thousands of people migrating to Sidney, Ohio. Don Abbott was one of them, and he remembers how he came to be employed for Ralph one day in 1950.

"I was out of a job, and so I went to apply [at a building that had been set up as a recruiting center]. There was a long line of people from Kentucky and Tennessee and almost everybody was hired. They would just bus them up to Sidney. But he didn't want any union workers. Ralph didn't care if the workers had any experience or not, but he didn't want union workers. So, anyway, I got through the line and told my story, and I was told by Mike Seaving: 'If we need you, we'll call you.' So I went back home, feeling pretty low, and I told my wife what happened. Well, we had a one-year-old daughter, and Hazel said, 'Well, one of

us has to work. If they won't hire you, they'll have to hire me.' She got the job because she was born in Tennessee. I didn't get the job, because I worked for a company called Monarch, and they had a union. Well, Hazel came back, with her job, and she told me: 'You can babysit.' Well, that did it. I said, 'No way, I'm getting a job.' And so I marched back to the line, and I told the guard, 'I have an appointment.' I went in and explained how Hazel had got a job, and I said, 'I need a job, too. Any job you have will do.' Well, Mike goes to Floyd Valentine's office, and they talk, and Mike comes back and he asks: "Can you go to work tonight?" And so I went back to Hazel and told her, 'Well, I got me a job, too.' And so we each worked and got a baby-sitter. Hazel worked there until 1970. I was there until 1986."

People who went to work for Ralph tended to stay until retirement, and many who left, came back. Jack Fisher, who started working for Ralph in 1959, left four different times to work at various companies in Indiana, and each time a job ended he returned to work for Ralph in Sidney. "I'm not particularly proud that I left," says Fisher. But he was able to come back so often because when you worked for Stolle, in a sense, you were working for your family.

And certainly Ralph did everything he could to foster a family environment at the Stolle Corporation and the Stolle Company, which eventually was an umbrella for at least twenty different companies within it. Rita Hauck started working for Ralph on May 20, 1953, and she worked for him up until his death, and one reason, arguably, was that he treated her better than most of his peers would have. In 1958, when Hauck's daughter, Melissa, was born, it looked like the end of a career. But Ralph appreciated Hauck's work—she was in accounting payables—and so he asked if she wanted to set up an office at home. She worked from her house until 1970, when she began working at the office again; for twelve years, she never saw Ralph, but spoke to him every couple of days on the telephone, and she frequently gave accounts and receivables to Ralph's driver, Slim, who shuttled the papers back to the headquarters. Decades before it became popular, Ralph understood the value of telecommuting.

Or he was just aware that it paid to be nice. Long before the movie *Jerry McGuire* popularized the idea of a mission statement, a 1985 mission statement titled "the Stolle Approach to Business" indicated the importance of having "good communications and sincere, trusting relationship with employees." Following that is a reference to a solid "quality of life."

Ralph tried to give his employees a strong quality of life, and he did his best to make his workers feel wanted. Weldon Oakley remembers ending one night shift during the Christmas season of 1956 and being introduced to his boss, who was in

a nearby kitchen fixing the workers an elaborate breakfast. As soon as Weldon referred to his employer as Mr. Stolle, he was told:

"Young man, remember, it's always Ralph. Just call me Ralph."

But what really made an impression on Weldon was the breakfast. When he went back to visit family and friends in Kentucky, he told everybody: "Hey, I've got a millionaire fixing me breakfast."

The millionaire would have gotten a kick out of hearing that story. As the fifth decade of the 20th century came to a close, Ralph was scoring points with just about everybody. Even his union problems had died down, so much so that in 1955, Ralph finally took a long vacation, not with his fishing buddies, but his family: He was able to take a rare three-week holiday with Dorothy, and the girls, Sandra, Gail, and Mary Jo. "It was a driving trip to Valley Forge, Gettysburg, New England, Montreal, and Niagara Falls," says Mary Jo, who was then a freshman in high school. "I couldn't believe that Daddy would take that much time off."

Probably, Ralph could hardly believe it himself, but the hard work that he had put into his companies since the 1920s was finally paying off. For years, of course, Ralph had had money, plenty of it, but that elusive quality of life was truly kicking in. Ralph had several successful businesses that spanned the state of Ohio; on the homestead, buffalo and cattle were wandering his farm. He had conquered his commuting woes by constructing his own airport. He had the resources to throw elaborate parties beyond anybody's wildest dreams. His protégé, Bob Mays, was growing up into a fine young man, more and more prepared to someday take over the business for him. By 1958, Ralph was even delving into science, having hired a university professor, Dr. Malik M. Sarwar, to conduct research at his dairy farm. Ralph had had a fascination with cows since he was a young boy, which is why he couldn't have just any old cattle roaming his farm: They were Holsteins, one of the most popular and finest breeds of cow, and a breed well-suited for milk production. And when he had read that a Dr. W.E. Peterson in Minnesota had made claims that cows could be induced to produce human antibodies, that was something Ralph immediately decided to investigate.

If Ralph hadn't conquered the world, he had at least seized a part of it, and to any outside observer, it must have looked like he was finished exploring any new frontiers. Clearly, this was a man who would bask in entrepreneurial glory for another several years, and then he could go through his retirement, wealthy and satisfied. Even Ralph might have agreed with that assessment at the time. Whatever his goals, he couldn't have known that a big part of his legacy lie just around the corner, waiting for him in the year 1962.

Chapter Five

THE MAN B

1959–63

"Cut the 'im' out of impossible . . ."

— NORMAN VINCENT PEALE

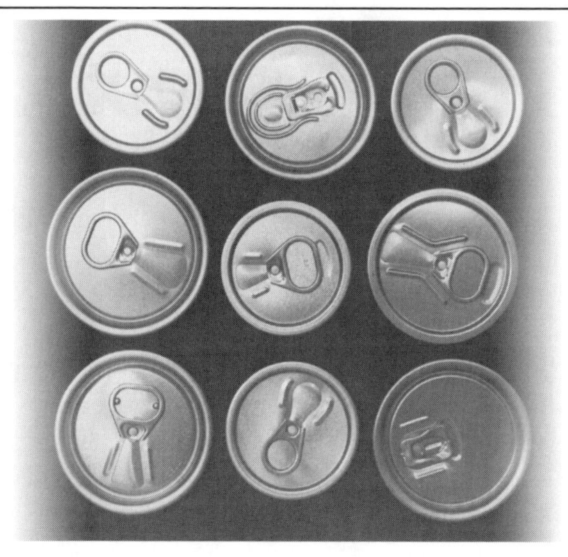

IT WAS STOLLE AND HIS COMPANY THAT INVENTED THE MANUFACTURING PROCESS THAT INCREASED PRODUCTION FROM 5,000 FLIP-TOP LIDS A DAY TO 1,000 PER MINUTE.

ERMAL FRAZE WASN'T THE FIRST to face this dilemma, but he made certain he would be one of the last. It was 1959 and the 46-year-old was on a family picnic with his wife, Martha and their two sons, Terry and Mark, as well as extended family members, when he realized he had packed cold cans of beer but nothing to open them.

Not a tragedy, of course, but with a lot of beer drinkers on hand, it probably seemed like one. It was also a frustration that everybody had run into at one time or another. If you had a can of soda or beer, your options for opening them were limited to a can opener, the kind with the triangular point at one end, which allowed a person to pry open a hole or two in the can, and these were usually referred to as a church key. The slang term was coined because the shape of the end of the opener reminded people of the ostentatious handles to large, old-fashioned door keys. And more often than not, these big keys opened church doors. But there was probably also some humor intended, since beer is a word you don't usually associate with church.

But praying is one of those words, and because Fraze couldn't count on divine intervention parting the metal on the cans, he opened them the only way he could figure out—by prying the cans open on the rear-end bumper of his car. It was a messy struggle between man and can, and he is said to have thought at the time—*There has to be a better way.*

But the better way didn't come until later, when one night he was lying awake, his mind wired from having drunk too much coffee before bed, and the idea just hit him: No one should be required to have a tool to open these cans. They should come with their *own* tool. Fraze toyed with his idea and ultimately obtained a patent for his invention, which was a removable pull tab opener for the tops of cans. In a nutshell, his invention was a rivet attached to a metal handle, which gave the user a fighting chance to open the can for drinking.

It's a terrific story, one that is repeated throughout newspaper articles and books of record, but it may well just be a story. A legend. Or it's simply the story without the rest of the facts.

It was a messy struggle between man and can, and he is said to have thought at the time—There has to be a better way.

In the early 1950s, according to Elton Kaminski, the United Shoe Machinery Corporation had a patent accepted for an easy open pull tab; the patent, which was the invention of one of their engineers, covered two different types of containers, one for food and one for beverages. While the company was generally known for its shoe-making machinery, it had over the years branched out into developing the hot glue gun, the drive mechanism for the lunar module and pop rivets used in the Supersonic Concorde. But while United Shoe had its patent, they couldn't develop a reliable way to attach the tab to the end of the lid of the can, and so they contacted Alcoa for assistance, explains Kaminski. One of the Alcoa branches, in Cleveland, was given the task of affixing the tab to the can lid, which brought John Hrovath, one of their engineers, to Sidney in the 1950s to talk things over with Kaminski and Ralph. But the two had to refuse Hrovath; they were simply too busy with other projects.

It was then that Hrovath met with Ermal Fraze. Certainly, Fraze had the background to develop the idea. While Fraze was born on a farm near Muncie, Indiana, he had an entrepreneurial spirit, selling newspapers and sandwiches at a canning jar factory and ultimately making more than the workers to whom he was selling. He also had a mechanical mind. After moving to Dayton, his first job was assembling the novelty prizes that came in boxes of Cracker Jack. And in 1949, thanks to an $800 loan from his young wife, Fraze had founded the Dayton Reliable Tool Company.

Fraze accepted Hrovath's assignment and, indeed, found a way to attach the tab to the lid of the can.

All of this, of course, doesn't mean that Fraze didn't once go on a picnic and find himself opening beer cans from his car bumper, or that he hadn't figured out how to attach the tab to the can during a sleepless, caffeinated night. But clearly, whenever the story has generally been told in print, some important details have been left out.

Fraze patented his work, which he allowed his own business to use, Dayton Reliable Tool Company, as well as the United Shoe Machinery Corporation. And that might have been it. Except that Alcoa was involved. They controlled the licensing of Fraze's patent.

. . . Close soon discovered that beer company executives had taken different math classes than he did, and that they would need some persuading.

And as far as Alcoa was concerned, without being able to mass-produce the creation, Fraze's patent was almost useless. Mass production was very much on the minds of Alcoa executives. The aluminum company had just eased into the aluminum canning industry in 1961, at a time when virtually all beverage cans were manufactured out of tinplate steel. That they would turn the canning industry on its end—no pun intended—was largely because of the ambitions of Fritz Close, the man who had climbed new heights during the lows of the Depression when he and Ralph joined forces to work on both the Empire State Building and Rockefeller Center.

As the decades passed, the 1930s becoming the 1940s, and the 1940s washing away into the 1950s, and finally the 1960s arrived, Close having risen through the Alcoa ranks from an aluminum salesman to vice-president of marketing. The years hadn't diminished Close's enthusiasm for aluminum. He was determined to see that beverage cans would no longer be made of tinplate steel. Close envisioned a day when aluminum would rule the canning industry.

Close had already scored a coup in the frozen juice and automotive oil market. Alcoa's nemesis, Reynolds, had been manufacturing aluminum juice cans and dominating the industry, but Alcoa trumped them by designing a foil and fiberboard container for frozen juice and—interestingly enough—automotive oil. It was a design that is the standard to this day.

But each year, there were billions and billions of tinplate steel soda and beer cans on the production lines. If Alcoa could grab even a tiny share of that market and produce aluminum soda and beer cans, it would mean great things for the already great company. And while Close had some interesting arguments that his metal was a better container for beverage companies than steel—for instance, some evidence existed suggesting that shelf life was extended when it was sealed in aluminum—he knew that if he could offer companies cans that came with an easy way to open them, he would have a far superior product that any tinplate steel can could ever offer.

If only Close could get them to see it his way.

The problem was that there were high hurdles to pass. Adding the easy open tab would make the beer can manufacturing process more expensive. Close had a ready answer for executives who would balk at the price—that Alcoa's marketing managers had run extensive studies showing that while the expense would climb, more beer would be sold if the tab was available, meaning that ultimately the cost would decrease. But Close soon discovered that beer company executives had taken different math classes than he did, and that they would need some persuading.

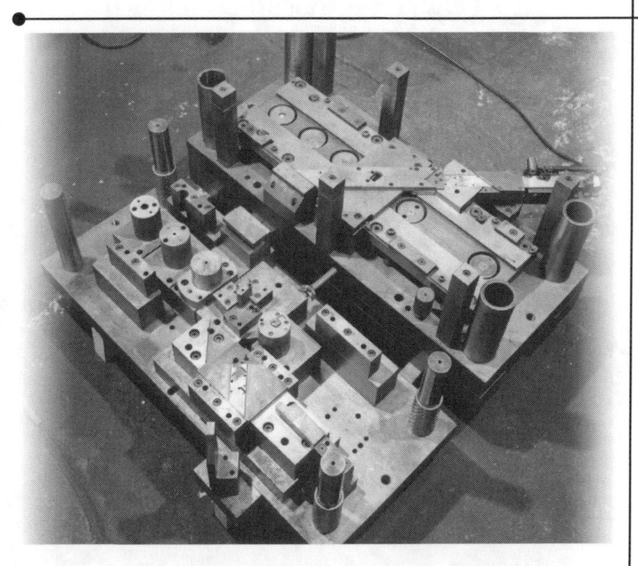

STILL AFFECTIONATELY CALLED THE "MINSTER MONSTER," THIS UNIT WAS THE FIRST TO PRODUCE EASY OPEN ENDS FOR THE 12-OUNCE BEER CAN. WHEN RETIRED FROM SERVICE IN MAY 1963, THE UNIT AND TOOLING HAD PRODUCED APPROXIMATELY 150 MILLION ENDS.

And so on one occasion, Close and his marketing manager, Cliff Sands, took some sample six-packs with aluminum ends made by their laboratories, and they traveled to St. Louis to meet with Augustus Busch, the president of Anheuser Busch. It was short meeting. When the price was put on the table, Close said his product would cost $2 more per thousand over the base price of steel, and Busch showed Close and Sands the door.

It may have been just as well, because there was this nagging problem. Ermal Fraze's design still lacked the components that would make it easy to mass-produce. Alcoa had a winning idea but no way to quickly churn out billions, or even millions, of the cans. At the time, the Dayton Reliable Tool Company was only producing 5,000 cans a day. And so one day, Close did what he had increasingly done over the last thirty years when his company had come against a dead end.

He picked up the phone. And he called Ralph.

Close asked the critical question. Will you be able to come up with a manufacturing method to mass-produce these easy open pull tabs? Fortunately, the answer was refreshingly familiar.

"Sure, I can," said Ralph.

It was suggested that Ralph pay a visit to Fraze's Perry Street, Dayton, location to see how things were going, and in early January, 1962, he did just that, bringing along Elton Kaminski, his chief engineer. For whatever reason, Fraze wasn't there to meet with them.

It didn't take long for Kaminski to come up with a solution, although carrying out the solution would take awhile.

. . . Close did what he had increasingly done over the last thirty years when his company had come against a dead end. He picked up the phone. And he called Ralph.

Fraze's machinery was producing the easy open pull tab one at a time, in what was called a rotary system. Kaminiski's idea was to develop what was an in-line transfer system, which "allowed high speed indexing movements through the tooling." Kaminski suggested that they consider the movie camera, which uses a flexible film, perforated on the edges, and takes pictures while indexing at sixteen frames per second. The film is then projected on a screen greatly enlarged, and the picture is perfectly registered. It was a strategy that Kaminski planned to follow, using aluminum.

Ralph saw the potential immediately. Which is not surprising. He had, after all, attended those two years at Ohio Mechanics Institute and had spent a lifetime working around machinery. He was a businessman through and through, but one who often understood the most minute inner workings of his company. Ralph gave Kaminski the go-ahead, and four months later they had a prototype machine that they converted out of an available press in Plant 1.

Meanwhile, Close had his own engineers working on developing the final design of the thickness and what the beer can lid itself would look like. And Close was hustling, trying to nail a big sale. Finally, he attracted the interest of a major player in the beer industry, Continental Can, which manufactured cans for Schlitz. In his meeting with Continental Can, Chase let it slip that he had been discussing the easy open tab idea with Augustus Busch, conveniently leaving out the part about him being booted out the door.

In the autumn of 1962, Continental Can placed a 200,000 order for the easy open can lids, to be attached to their beer cans. Close had to be ecstatic, as well as a little terrified.

Because there was this nagging problem.

The Stolle Corporation wasn't quite finished with the machinery that would mass-produce these easy open lids. As Kaminski now observes, "The rush... was on."

Fritz Close was nervous. He didn't want to just discuss how things were going over the telephone, or via letters, though they did talk and correspond. He wanted a personal meeting with Ralph, who was only too happy to oblige.

Ralph sent his plane out to Pittsburgh to pick up Close, to give his associate and friend the royal treatment. Fritz was flown to Sidney, and he traveled back to Lebanon with Slim Buchanan at the wheel; his passenger-mates were Ralph and Ralph's attorney, Nelson Schwab, who over forty years later remembers fragments of the discussion. That Ralph was still exuberantly optimistic that his company could invent machinery to mass-produce these easy open tabs. And that Schwab, at one point, said: "I can't tell you how much that will add to Alcoa's appeal. I wish I could stop the car right now so I could get out and dance a jig."

Upon reaching the ranch house, Ralph prepared the meal. Kaminski was also present, as was Cliff Sands, Close's marketing man, and Don Welty, a retired Alcoa research and development director. This was weighty stuff, but Close was delighted with everything he heard. The vice-president of marketing flew back home and immediately had a letter typed up.

"Dear Ralph," the letter began, "I want to take this opportunity to thank you for last evening. In true Stolle fashion, the dinner was delicious. As you know, Ralph, I have made many trips to not only your plant, but your home, but I have never been on any more serious mission than I was last night."

Close went on to tell Ralph that "the complete assurance" they were able to give him—that everything would be finished on time—"was the best news I have heard in a long time," and further in the letter, Fritz revealed his anxiety when he stressed: "As you know, Ralph, we and you are in the same sort of situation as a fellow jumping with a parachute. It's got to work the first time."

Close wrapped up the missive by letting Ralph know that Continental Can would be sending them 200,000 ends of the cans for them to fasten the easy open tabs, and that they would be at the Stolle Corporation by November 1. Close's letter was dated October 12. Clearly, Close was in a dinghy being swept down the rapids, and Ralph had just agreed to take over as captain, fully understanding that he and his company's reputation might end up going over the falls if things didn't work out just as planned.

It's a romantic notion to think that the entire Stolle operation went into full gear, with each employee working around the clock, sleeping at their desks, bleary-eyed at the machines, all in the name of getting the easy open end tab finished in time. But actually, there were perhaps only several dozen people whose lives were thrown into overtime turmoil, most of them people who worked at the Sidney Machine Tool Company, a business bought up by Ralph to assist in the easy open end tabs, and where much of the work would be done. Elsewhere, most of the Stolle Corporation employees had little knowledge of what was happening, and most of them who knew, just knew it was something big.

THIS WAS THE HIGH SPEED TRANSFER PRESS, ONE OF THE CRUCIAL COMPONENTS IN PROCESSING THE EASY OPEN END CAN LIDS.

Ralph's secret weapon was Elton G. Kaminski, who longtime Stolle employee Bob Mays says "was the brains of the [easy open end] operation." Colleague and former Stolle Corporation president Dick Pope describes Kaminski as a reclusive genius. "I think he found it hard to relate to people," says Pope, "but he was brilliant, constantly coming up with ideas." Kaminski had grown up in Covington, Kentucky, and was a graduate of Northwestern University, an engineer who served as a soldier in Germany at the end of World War II, before finding a home at the Stolle Corporation in 1951. And home is an apt way to put it.

By 1951, Kaminski was already a third-generation Kaminski working for Ralph, and he had first met the CEO when he was 9 years old. Both his father and grandfather, Herman and Frederick respectively, were Stolle employees in Cincinnati in the early 1930s. Frederick was getting on in years and wouldn't stay with Stolle for long, but Herman started as a maintenance worker and was the company's master mechanic when he died in 1948. Herman showed much of the same brilliance that Elton possessed, according to Tom Anderson, who worked there for awhile as a teenager in

1943. "Elton's dad was head maintenance man and quite a character. He was almost a genius. Ralph could make a diagram of what he wanted, and Herman could build it. He was a heavy drinker, but it didn't deter him from his work."

Ultimately, the Kaminiskis ended up having their own family company within the Stolle Corporation. Ralph eventually hired Elton's daughter, Tina, to work in computer graphics, and his brother, Don, who for many years held various positions in maintenance and engineering.

But as important as Kaminski was to the success of the easy open pull tab, everybody—including Kaminski—is quick to credit Ralph for their success. Roger Brautigam puts it this way: "He gathered people like you gather apples… Mr. Stolle had the knack and theory in his head, that if you put a group of dedicated people together, with the right tools, they can do almost anything."

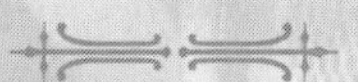

Bill Littlefield remembers Charlie Williams, his foreman, explaining to him and about forty other employees that they were about to embark on something big. For the last ten years, Littlefield had been manning the pantograph, a machine that was used to cut the score, which is the part of the apparatus that creates the indentations in the can lid. Before the can lids came along, Littlefield and his co-workers were making trim for automobiles, ovens, and refrigerators. Those responsibilities would ultimately be switched to another part of the company. Littlefield was about to help make the lids of beer cans.

Trouble was, of course, nobody quite knew how to do this. Littlefield and his crew had to deduce exactly how much aluminum could be in the can, and how thick the width would be. "I've worked with metal that's 1/1000th wide, but we were working with metal 1/10,0000th wide," marvels Littlefield.

What seems so simple today would have been a mechanical miracle back then.

Littlefield says that the higher-ups made it clear "the situation was urgent." He logged in a lot of overtime, going from working eight- to ten-hour days five days a week to working twelve- to sixteen-hour days five days a week. And then they started working some Saturdays, and occasionally, Sundays. "I remember working eighty-five hours one week, and I know another fellow who worked over a hundred," says Littlefield, whose wife complained, but not much. "She knew I had to do that, that it was part of the job." And with four children to feed, the Littlefields were grateful to have the extra overtime income.

Roger Brautigam didn't mind it either. "In the tool and die business, there is no slack time," he says matter-of-factly. "If you had a die to build, they expected you to be there." He says that the workers logged in twelve-hour days during the week, ten hours on Saturdays and a half a day on Sundays, for a long stretch of time. "Our families realized we had to do that." It was a pressure-filled time. "He expected a lot of us," says Brautigam of Ralph, "and we knew it, and if you want to perform for somebody like that, you don't lean back and stick your feet on the bench. You're into it."

But Brautigam is adamant that he isn't complaining about the time spent on the pull tab, or that one gets the impression Ralph was a hard task master. You end up with the idea that Brautigam would have taken a bullet for Ralph, if asked.

Meanwhile, at the ranch...

Ralph's daughters remember seeing beer can lids piling up in the living room, the kitchen, the bedroom, the hallways, the basement... They were used to seeing strange things throughout the house, such as Whirlpool insignias, but the sight of all of the numerous can lids had become kind of amusing, and surreal. "The big problem that I remember was getting these cans not to leak," says Sandy. "I remember them struggling with that."

"There were cans all over the house," agrees Mary Jo.

Slim had been making deliveries to anywhere from Detroit to Dayton since taking his job, but after driving Close to Ralph's ranch, there had been numerous reasons to keep him behind the wheel more than usual. For about six months, all he seemed to do was drive, recalls Slim. One Saturday night, Slim found himself pulling into the parking lot at the Midwest Tool & Dye in Dayton. It was 1:30 a.m. Slim was here to pick up the prototype of the can lid with the easy open pull tab. Surely the hours were going to be getting better soon.

Slim met with a weary employee, who handed him the prototype. Which then slipped from Slim's hands.

If this had been a scene in a movie, the prototype would have fallen in slow motion, as the camera panned to the look of terror on the two men's faces. Instead, it only took a second for the prototype to fall to the floor.

And even less time to break.

"I thought I had had it," Slim would say, some forty years later.

The exhausted employee, however, picked up the broken prototype and retreated back into his workspace. Three hours later, he returned to hand a relieved Slim the prototype. Slim cradled it gently until he reached the car. And then he stomped on the gas, making a beeline to the Stolle Corporation.

As the deadline was approaching, Kaminski's ideas were working, but it wasn't a flawless system. The tabs were fragile. They would fall off. They wouldn't go through the machinery properly. Kaminski knew what the problem was, but to fix it, the can companies needed to make some adjustments on their ends.

"Their plain can ends required a lubricant on the surface to allow proper formation of the rivet that held the tab to the can end," Kaminski would write years later. The problem? "We had an impossible time convincing the canmakers."61

In what was a seminal moment in the aluminum can industry, and certain the history of beverage drinks, Ralph had an idea on how to convince the stubborn canmakers that the ends supplied to the Stolle Corporation needed lubricant. Ralph sent Slim out to a local market to purchase some canning paraffin, and then in his kitchen facilities, Ralph melted the wax in a pan. It didn't take long. Once it was oozing in the pan, Ralph dipped some can ends in the paraffin, and he had them rushed to Plant 1 where they ran them through the machinery, along with some can ends provided by the canmakers. Kaminski was vindicated.

"After passing through the easy open end conversion presses," he writes, "the ends dipped by Ralph exhibited perfect rivet formation while those not dipped showed misshapen, unreliable rivets and caused premature wear on the rivet- forming tools. The canmakers were convinced and rushed to provide the required lubricant on the ends, which they furnished us.

"Today," adds Kaminski, "the necessary lubricant is incorporated into the enamel coatings applied to the ends before they are formed by the canmaker."

The Stolle Corporation delivered beyond Close's wildest expectations. In 1963, the operation went into a twenty-four-hour, seven-day-a-week production, and continued at that pace for over nine months. The idea was that they would produce a tremendous supply of lids to keep everybody from running out of stock while the buyers of the lid-making machines had their equipment moved from Sidney to their own plant locations. By the end of 1963, the aluminum lid with the easy open pull tab was being used on forty percent of all American beer cans. By 1968, cans with aluminum ends had more than eighty percent of the beer market. Meanwhile, the added business for Alcoa wasn't hurting the Stolle Corporation. While it's not generally known how many millions of dollars were flooding the company, their success could easily be seen by their expansion; for the twenty-four-hour odyssey, Ralph purchased the old Sidney Machine Tool Company facility on Highland Avenue, and then later he opened a new 400,000-square foot plant on Campbell Road for residential aluminum siding and its accessories. And the Stolle

Corporation began buying up companies at a fairly rapid clip, starting with, in 1963, the Norcold Refrigeration Company. By the mid-1970s, the Stolle Corporation would have nine plants scattered across Sidney, producing easy open pull tabs, creating decorative trim for automobiles and appliances, fashioning aluminum siding, gutters, windows, shingles, and commercial building panels.

Meanwhile, the Ralph J. Stolle Company was expanding, too. By the late 1970s, some of the companies that Ralph owned included: the Sheffer Corporation in Cincinnati, a manufacturer of hydraulic and pneumatic cylinders; the Deerfield Manufacturing Company in Mason, Ohio, which made heavy steel stampings and metal assemblies; the Imperial Tool and Die Company, also in Mason, which made tools and dies; Lebanon Ford Tractor Company, in Lebanon, Ohio, which sold and serviced farm implements; and the Sidney Elevator Manufacturing Company, which manufactured sheet metal products and high accurate tooling. Ralph even owned the Stolle Realty Company, which dealt with a lot of farmland.

These weren't all glamorous products and services, but Ralph's fingerprints were touching products around the world. And there must have been some satisfaction for Ralph that his always money-conscious mother, Kunigunde, lived long enough to see Ralph reach the big-time. She passed away on April 19, 1965, at the age of 89.

Indeed, the pull tabs were the big-time. More than $100 million was invested by Alcoa at its new smelter in Warrick, Indiana; from around the country, barges of alumina, a white granular material that would later be transformed into aluminum, were sent down the Ohio River, past where the fledging Stolle Steel & Iron Company once stood, and to the smelter in Warrick. And around the world, the easy open pull tab was catching on. In 1967, the soda pop giants, Coca-Cola and Pepsi, adopted the easy open pull tab, and there was just no going back. Everybody recognized the power of the pull tab. It even inspired a laughable lawsuit; two convicts filed suit against a beer company, Iron City Beer, in Pittsburgh, claiming that it stole their easy open pull tab idea after an in-prison interview sometime before June 9, 1961. The convicts would have been better off suing Ralph Stolle, Fritz Close, Ermal Fraze, Elton Kaminski, or somebody who actually had a hand in creating the pull tab.

But there was an ominous undercurrent with the pull tab phenomenon, something that nobody had counted on: human nature. Unfortunately, not every one is predisposed to throwing away garbage. The invention had worked too well. The cans were a little *too* easy to open. As the 1970s arrived, the planet was becoming littered with millions of aluminum pull tabs. People were cutting their bare feet on tabs found on beaches, and animals were choking on them. It was enough of a hazard that there was even some talk that states would start banning the aluminum can. That led to refinements in the can design, much of which was done by what would be known as the Stolle Machinery Division, and ultimately cans became

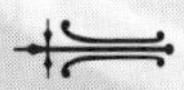

what they are today: The pull tabs are still easy to open, just not as easy to pull off.

There was another ominous residue from the pull tabs, once again due to human nature. Somehow in the early 1970s, an urban legend started making its way around the country that if you collected enough pull tabs, you could give them to the National Kidney Foundation, so they could recycle them, and they would give a patient free time on a dialysis machine. As recently as 2002, in Birmingham, Alabama, a girl with melanoma had been duped into believing that Coca-Cola would pay for her immunotherapy treatments, giving her one free treatment for every 1,000 pull tabs collected. After the girl's friends, family and school had collected 276,000 of them, the girl's mother called Coca-Cola and learned the sad truth. If Ralph had been around to hear about it, it's a good bet that he would have paid for the immunotherapy treatments himself.

But some good has come from the urban legend. In 1987, after years of seeing people collecting pull tabs for no avail, the Ronald McDonald House in Minneapolis and St. Paul designed, "If you can't beat them, join them." And so they created the Ronald McDonald House Pop Tab Collection Program, which was later adopted by other Ronald McDonald Houses across the country. To help their cause of offering cheap lodging ($10 a night) to families who have children in hospitals for extended stays, the Pop Tab Collection Program has collected over 400 million pop tabs, raising over $300,000.

And in 1989, shortly after the man who started it all, Ermal Fraze, passed away, Dayton was so proud of his accomplishment that they named a popular amphitheater after him, the Fraze Pavilion.

PROTOTYPE OF ANOTHER CAN: STOLLE ALSO DEVELOPED A SYSTEM TO PRODUCE, AT HIGHER SPEEDS (USING SIMPLER EQUIPMENT), SHALLOW AND DEEP DRAWN CONTAINERS.

And so Ermal Fraze and Ralph Stolle, both cut from the same entrepreneurial cloth, would be forever linked as the men who created the easy open end pop top tab. They would never be more than acquaintances, and not really that—they met only once, briefly. But because of their individual roles in arguably one of the most utilized inventions of the 20th century, when Ermal and Ralph's respective times came, the obituary writers ladled out grateful and kind words for both of them.

1960s & ... GOOD

"Always do right—this will gratify some and astonish the rest."

— MARK TWAIN

THROUGHOUT HIS LIFE, RALPH SENT 80 OF HIS EMPLOYEES' CHILDREN TO COLLEGE. PICTURED ABOVE IS A GOOD AND CONFIDENT MAN IN 1975.

THERE WERE PROBABLY TIMES when Mike Seaving felt that he had the worst job in the world, although most days, he believed it was among the best.

The funerals were emotionally draining. Whenever anybody died—a Stolle employee, a spouse, the son or daughter of an employee—Seaving was at the funeral, because he felt that's what you did when you were the human resources director. From the moment Seaving started working for Ralph shortly after World War II, he had attended all of the funerals of anybody who worked at the Stolle Corporation. He shook the hands of the grieving widows, slipping $20 into their hands as a way of letting them know they would be taken care of. He made the funeral arrangements when he had to, and he told shell-shocked husbands and wives about the company's generous life insurance plan. Then he drove away in his white Ford Thunderbird.

Uneasy people still had their shattered lives to repair, but after a conversation with Mike Seaving they usually felt better, and often extremely grateful, that they worked for such a man as Ralph Stolle.

And then there were the times that Seaving came to the rescue of a Stolle employee, not with a kind word and a hopeful message, but by peeling paper bills out of a wallet and greasing the palms of an anxious landlord to get a worker's apartment back. Or Seaving would be there with bail money to free an employee from jail, if one drink too many had sent them into the clink. As Seaving was a distinguished looking gentleman with silver hair and a suit and tie, he always looked a little out of place when he appeared at the police station.

But he was no pushover. He could handle the toughest of the Stolle employees. Like Ralph, you always got the sense that you shouldn't cross Seaving, who was an imposing height of at least six feet.

Every Monday morning, the foremen told Seaving who hadn't shown up for work. If trouble was suspected, Seaving called that employee's home. With 1,700 workers to watch over, there was almost always somebody who wouldn't show up, and every once in awhile, it was because of a problem.

If there was a problem, Seaving had one mission: fix it, and, if feasible, get the employee back to work. "Whatever it took, no questions asked," says Ralph's best friend, George Henkle, who heard many stories over the years, most of them coming through not the reticent Ralph, but their circle of friends.

Through Seaving, "Ralph took care of his employees like a father figure might, even if the situation was their fault. Now, the repeat offenders, he wouldn't help. Those guys, he eliminated one way or the other," says Henkle, presumably not meaning to sound as though his friend would have fit right in with *The Sopranos* or one of the *Godfather* movies. "To Ralph, the worst thing you could do was gamble," says Henkle. "It was worse than drinking, or probably any other habit, because you were usually gambling the grocery money and taking food out of the mouths of your children. And so a gambler couldn't expect much help, if any. But if it was a sick wife who needed medicine, or if you couldn't pay for groceries, or your car was in an accident and needed repairs, or you were in jail, or whatever, Ralph didn't ask any questions. He just had the problem fixed. That's why he had so many loyal workers."

Ralph didn't ask any questions. He just had the problem fixed. That's why he had so many loyal workers.

But it was Seaving who did the unenviable work of visiting jail cells and funeral parlors. However, he always did what he did in Ralph's name. "He was Ralph's right hand man," recalls Dick Pope, who got to know Seaving well. One day, in the late 1960s, Ralph told Pope that he would someday have Seaving's job, and that Seaving would mentor him. Pope was in his mid-20s; Seaving was approaching 70. Not only did the job not interest Pope, hanging around Seaving didn't seem too appealing either. As distinguished as Seaving could appear, he had contradictory eccentricities. He could talk for hours without seeming to say anything important. He had no teeth and never did, as long as Pope had known him. His body was racked with diabetes, which didn't stop him from smoking continuously. Seaving had a habit of walking around the plant and slipping his hand into his pocket, pulling out a cigarette from the box without removing the box from the pocket. "It's not easy to do," says Pope. "Believe me, I've tried."

Before Ralph's mandate, Seaving had nonetheless been hanging around Pope often, talking to him for what seemed like hours at a time. Later, Pope realized that Seaving was selecting him to take over the human resources position, with Ralph's blessings.

> *. . . Ralph recognized his employees' worth to the company, and that without his employees there would be no company.*

After Ralph's mandate, Pope and Seaving spent what no longer *seemed* like hours together—it *was* hours together; and slowly, the young Pope began to learn what both Seaving and Ralph had an uncanny understanding of: people. "Mike had a lot of theories about human behavior and why people act the way they do," says Pope, who nevertheless didn't appreciate everything Seaving had to say. And Pope shared his frustrations with Ralph.

"I've had no training, I'm not cut out for this, I don't know what Mike is trying to tell me," began Pope one day.

"I don't care," Ralph interrupted. "You have to learn everything he knows. And it won't be easy."

Pope flushed, his temper rising. "This isn't what I want to do—I want to do something else."

Ralph countered. "Look, I know that seventy-five percent of what Mike says is bull crap, but the other twenty-five percent is jewelry. You have to decide what the jewels are."

Somehow, Pope suddenly felt better. At least Ralph seemed to understand where he was coming from. Pope continued meeting with Seaving and years later marveled at what he had ultimately learned. "What the hell," decided Pope. "That old man was pretty smart."

Pope learned the importance of getting to know everybody, inside the company and out. That Seaving knew everybody in Sidney—*everybody*—meant it was easier to twist arms when you had to. Seaving could make things happen. Both Seaving and Ralph, and it's hard to know who influenced whom, fervently believed that a happy worker was a productive worker, and that employees liked to feel appreciated and part of the team. Which is why Ralph always had Seaving at his side, whenever touring a plant. Seaving knew all of the names of the employees, and the names of their family members, and probably the names of their friends; it was his job. Just before greeting anybody, Seaving would whisper the man's name into his boss's ear, and maybe a little information about them, and Ralph would strike up a conversation. That the boss knew everybody's name made him a legend among his staff. And if the familiarity among his staff was really a hall of mirrors, at least Ralph cared enough to make it clear that he *wanted* to know everybody's name.

STOLLE (SECOND FROM LEFT) AND JAMES A. RHODES (SECOND FROM RIGHT) WERE BEST FRIENDS. RHODES WAS GOVERNOR OF OHIO FROM 1963 TO 1971, AND THEN FROM 1975 TO 1983.

Meanwhile, Pope started memorizing names and faces and trying to keep them all straight.

Then one day, Pope was on vacation when Seaving called and said he was coming over. They ended up sitting in front of Pope's house, chit-chatting in the Thunderbird, when Seaving gave him the news: "I'm not coming back to work."

Pope was incredulous. "Why?"

"Well, it's time for you to take over."

And true to his word, Seaving didn't return to work, though he immediately began telephoning Pope to see how he was doing. Two weeks later, Seaving had a heart attack and was dead. Pope attended his funeral, but he didn't slip Seaving's grieving widow a twenty. She was well-off and may have not appreciated the gesture. It just didn't feel right.

Of course, many employers might have just let their employees and their families fend for themselves, but Henkle theorizes that one reason Ralph was so principled to his workers was because of his Southern upbringing. "His mother and father taught him to do that. It was the culture he grew up in."

However he came about his ethic, Ralph recognized his employees' worth to the company, and that without his employees there would be no company. As long-time employee Maxine Pence says, "Ralph knew that if your family was in good shape, the workers were in good shape."

Sunday nights, some of the workers weren't in good shape. Monday nights, they were.

Slim remembers it like it was yesterday: a 310 Cessna appearing in the sky, circling the lonely roads leading to Sidney. Slim watched the plane come in from a grassy field; today, that field is a Bob Evans and its parking lot. Folkers Airport is gone; in its place are restaurants, gas stations, and a grocery store. The nearby hotel, the Wagner House, is now a FifthThird Bank. But except for the hotel and airport, there was precious little around, except for Slim, who waited with his 1962 Chevrolet Mercury, as Dan Phelps—heavyset and a chain-smoker—landed on the grass runway.

RALPH SENT HIS AIRPLANE AROUND THE COUNTRY, AFTER EMPLOYEES WHO BECAME ILL ON VACATION, OR WHOSE FAMILY MEMBERS HAD PASSED AWAY. HERE HE IS WITH CO-WORKERS, DEPARTING ON A BUSINESS TRIP.

These were good days for Ralph, who had bought his own airplane sometime ago, and hired his own pilot to fly it. He would spend nights in the suites at the Wagner House, without ever caring about the cost. It simply wasn't an issue for him. Money hadn't been much of a problem for a few decades, and it would certainly never be again. Not after the success of the easy-open pop-top tabs, which were rapidly appearing on cans across the world.

While Ralph probably could have wallpapered the hallways of his ranch with $1,000 bills, despite his company's stratospheric leap into success and wealth, Ralph was careful not to flaunt what he had. Everybody from Slim Buchanan to Ralph's granddaughter, Cathy Chasteen, has marveled at how

this millionaire embraced simplicity. Ed Cranmer recalls that his grandfather dressed so modestly that "if you saw him on a street, you wouldn't suspect he had even a quarter in his pocket."

To a point, this is completely true. Ralph wasn't one to go for fancy cars or European suits. He didn't lord it over anybody that he had money.

On the other hand—Ralph was rich, and he didn't exactly *hide* the fact. He owned a farm that stretched out for miles, and he raised numerous exotic animals, including ostriches and baboons. He had a herd of bison roaming on his ranch, eight cows and two bulls, the only place in Ohio where you could find such an animal, a Cincinnati newspaper reported in 1950, going on to say that schools were arranging visitations for children, and that traffic was slowing ever since the animals had arrived. Ralph's airport was just up the hill from his ranch house, complete with a runway and hanger for his airplane. He threw lavish parties for the closest hundred of his family and friends. The parties, in fact, were so big and crowded, often people were wearing nametags. Ralph never let anybody pay for dinner when he was out at a restaurant—even in the company of thirty or forty friends. He didn't go to a football game with a few pals; he rented a bus from a buddy, the Ohio Secretary of Transportation, and he took along several dozen people.

[This] all seemed to fit into Ralph's philosophy: Don't hoard the gold; share it.

Which all seemed to fit into Ralph's philosophy: Don't hoard the gold; share it.

The gold that Ralph gave away is legendary around the communities he lived, among the people who remember him. In his 92 years on the Earth, Ralph sent eighty of his employees' children to college. And for some of them—he didn't just send them to college—he also paid for those students to continue onto medical school. All he asked in return was that when that person was rich, they help somebody else go to college. Long before the expression "pay it forward" was used, Ralph was asking people to do just that.

There's one story floating around that Ralph once fired a worker, and then he got him back on his feet by helping him start a restaurant. Meanwhile, Henkle recalls a business associate who stole material from a lumber company that he had started with Ralph; the material was wood, of course, and enough to build an apartment complex that the associate was trying to secretly do on the side. Once Ralph learned about the theft, he promptly dissolved the business, but he forgave the man and was friendly to him whenever they came across each other in the same social circles.

Ralph sent his airplane around the country, after employees who were ill on vacations, or whose family members had passed away, and he generously gave away a mint of his money—likely well into the millions—to churches, schools, YMCAs. And he, in a sense, offered his land to live on rent-free to people he simply liked. After his death, his estate discovered that Ralph had forty-two farms other than his own scattered across the country, and many of them had tenants whose only job was to apparently mow the grass. "Ralph said I could live here as long as I live," was the common refrain when they were contacted.

Dan Phelps, the chain-smoking pilot, didn't have health insurance, which was something he sorely missed when it was discovered that he had lung cancer. Ralph had Phelps taken to Chicago for experimental treatments, but the doctors couldn't save him, and within the year, Phelps passed away. But during the year that Phelps was sick, and for about a year after that, until his widow moved her family away, Ralph would visit the household and give Phelp's wife $200 in cash.

Whenever he could, Ralph clearly relished playing a modern-day, real-life superhero, helping the little guy and often dashing away before he could be thanked.

Art Middleton owned a small grocery store before becoming a twenty-year maintenance man at the Stolle Corporation. He remembers that throughout the 1960s, he would be called to the telephone by Slim, who always sounded urgent: "Ralph's found somebody hungry again. He wants you to pack up some bags."

And so Middleton shuttled through his store, putting together a few bags of staples. Slim would come tearing into the parking lot, Middleton handed him the bags, and off went the Chevrolet.

Middleton recalls that he wasn't the only merchant in town who Ralph utilized to come to the aid of another. Middleton says that under his boss's orders, Slim once dumped a load of coal in man's front yard—because Ralph had heard he had no heat.

When Slim wasn't racing to Middleton's, he was delivering food to a shelter in Sidney. Whenever the Stolle Corporation had an overabundance of leftovers, from Ralph's cooking or from a bash he had had for business guests, it was up to Slim to deliver it. "Ralph always insisted that I make the donations anonymously," says

Slim, who recalls that one time he gave away what must have been several dozen turkeys. Chances are, Slim admits, that the shelter knew exactly who was doing the donating, because everybody around town knew who Slim worked for. But nobody else knew, which is the way Ralph liked it.

Not too many years later, Dalton Messersmith and his wife, Linda, were both concerned for their son, Michael, who had cerebral palsy and made his way on wooden crutches. Like most boys, Michael liked to run, and when he did, the crutches always snapped in half, and Dalton's wallet took a hit. Word got around to Ralph that there was an employee with a son whose crutches kept splintering. And so Ralph issued an order to some workers in a metal shop to make Michael some aluminum crutches. "If he outgrows them, or needs new ones, get them here," Ralph instructed Dalton, and it was made clear that the Messersmiths had a lifelong supply of free crutches whenever they needed them. Michael is now in his 40s and in a wheelchair, but Linda says that they will never forget what Ralph did for their boy.

Ralph wasn't always the superhero, of course. As his nephew, Ron Pendery observes: "You don't get through a lifetime of business without rolling over somebody." There were some family members who clashed with Ralph, or maybe Ralph clashed with them. Ron's brother Ken adds, as diplomatically as he can, "There were members of the family he could have done without, who worked for him." Ralph was vocal in making it clear he would like to fire two people in particular, and Ken and Ron spoke up for them. "Ralph didn't know them like we did," says Ken. "Now, maybe there was something Ralph knew that we didn't, but we didn't think so, and we defended them."

> *"This man is so successful in business that he doesn't know how many millions he is worth, but he walks back to the hall with us, just so he can hold the door open for us. That is my big boss!"*

Ken and Ron both say that their uncle never held it against them for sometimes taking a side that didn't happen to be the one that Ralph was standing on. Because Ralph's business was a family business, perhaps it was inevitable that there were always some relatives who felt that their patriarch took more of an interest in people he didn't know well, than the people he did.

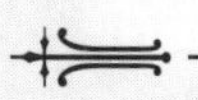

But what those relatives might have failed to realize was that for Ralph, his family business was truly a family business. All his employees *were* his family—a large, extended family, yes, but family all the same.

One telling example is a story that current Stolle Company president Bill Falknor shares:

"One of Ralph's employees—everybody called him Red—had a drinking problem. Well, Ralph never fired anybody, and so even though Red's drinking interfered with his work, Ralph kept him on, and he gave him the position of a guard at Plant Six. Well, one day, Red's really drunk, and Ralph told Seaving, 'Mike, I've finally had it. I want him out of here. He's never going to work for the Stolle Corporation again. Get rid of him.'

"Mike's answer? 'Done,' he said.

"Weeks later, Mike Seaving came out to the ranch, and who should he see, but Red, on a ladder, painting the house. Mike just stared at Ralph, who said, 'Dammit, all I said was, he'd never work at the Stolle Corporation again!'"

Twice a year, Ralph's extended family has a reunion of some sorts—in the form of a retirement luncheon at the Veterans Foreign Post facility in Sidney. New and old retirees come to reminisce and renew old friendships—and everybody seems more than willing to sing the praises of their old boss. "When Ralph was running the company, you were never laid off. You always had a job," says Don Boyer, who worked at the company for forty-one years. Actually, there were people laid off throughout the years, but it was so far and few between that nobody seems to remember any lay-offs—a job at the Stolle Corporation represented security, which is a concept that few employees around the country have had since the 1970s. From the 1930s up until Ralph sold his company to Alcoa in 1974, the message to the employees was always clear: Work for Ralph, and he'll look out for you.

The love and admiration that Ralph's retired employees have for him is palpable. Richard Borchers, a former Stolle employee, has some stories. He recalls how when Ralph was 80 years old, there were some sample parts to a product they were making, and the company couldn't get any polishers to come in on a weekend. So Ralph came in on a Saturday morning, wearing coveralls, and he helped the foreman polish the parts they needed. The foreman later told Borchers that he was embarrassed throughout the polishing, because Ralph was quicker at it than he was. And then Ralph didn't help by saying, "It's OK, just do what you can."

Another memory that Borchers enjoys is of some of the office people who, on a regular basis, came into the building late. But that stopped when Ralph came to work early for awhile, sitting in the lobby so he could greet each one.

But Borchers will never forget how the Stolle Corporation gave him and his wife's daughter, Janice, a scholarship to college. During a recognition banquet for the students, Ralph left early, heading into the parking lot. Borchers and his wife hurried outside to thank Ralph, who then walked back to the banquet hall, just to hold the door open for them. Borchers turned to his wife and said, "How do you like that for class? This man is so successful in business that he doesn't know how many millions he is worth, but he walks back to the hall with us, just so he can hold the door open for us. That is my *big boss*!"

Norman Jones is succinct when summing up his feelings about his old employer: "Ralph was a great man to my family and myself, and to many families."

LIKE A LOT OF BUSINESSMEN, RALPH NEVER STOPPED WORKING— EVEN ON THE GOLF COURSE.

It is a nice enough sentiment, but rather boring on the surface. After all, everybody at the luncheon is saying that Ralph is great. He's great, great, great and great. Which tells us little. And so when Jones, who worked at Stolle for almost fifty years, was asked what made Ralph so great, the burly man suddenly struggled to retain his composure.

"My son died in Vietnam," a pained Jones managed to say, and from the ashen look on his face, well over thirty years later, it's obvious that not a day hasn't gone by since 1969 that this father hasn't thought of his 21-year-old son, Monte. Norman Jones mumbles something. He's almost inaudible, but he refers to a collection, or perhaps a foundation, something in his son's name, which apparently was initiated or at least whole-heartedly endorsed by Ralph himself. And it isn't difficult to imagine that Mike Seaving, in Ralph's name, did whatever he could to make the Jones' family's suffering a little less so.

Jones adds that Monte once worked for Ralph as well.

But the father can't talk about it. Satisfied that he's made his point, Jones shuffles away, though not before repeating: "Ralph was a great man.

Chapter Seven

1970s LAND OF M

"Got milk?."

— POPULAR SLOGAN OF THE CALIFORNIA MILK PROCESSOR BOARD

IN 1968, RALPH SOLD HIS PRIZED HOLSTEIN BULL TO CARNATION FOR $250,000. EVEN GOVERNOR JAMES RHODES CAME BY FOR THE OFFICIAL PHOTOGRAPH. (ON LEFT, RALPH AND GOVERNOR RHODES.)

IN 1972, DR. LEE BECK WAS ASKED the first of two questions that would change the course of his life. But the question offered no such hints, and it seemed innocuous at the time: Would you be willing to meet Ralph Stolle in Sidney, Ohio, and talk to him about your research?

Beck didn't hesitate—much. He would go.

After all, the question came from Beck's boss, Vern Stevens, the chairman of the department of obstetrics and gynecology at The Ohio State University. And you just didn't say no to the chairman of your department, especially when you were a lowly graduate student working on your post-doctorate degree. Not when Ralph Stolle was a personal friend of your chairman's.

Besides, Stevens had chatted about Stolle enough for Beck to know the man sounded interesting enough. He might have become a millionaire in industries that were miles away from his own research, and he was well over twice Beck's age, but the two men had some common interests uncommonly found in others: They were each interested in immunization research.

But driving to Sidney and back would a take at least two hours of the day, not counting the time spent on the meeting, Beck reasoned, and it wasn't as though he had time to spare. The 27-year-old was toiling in the lab at OSU, working on his post-doctorate, and once he had that degree, he could start his new job that had been promised him at the University of Alabama. He needed a position, and one that was brimming with promise, with Margie staying at home, taking care of their 3-year-old son, John, and their newborn, Jessica.

There might be an upside to all of this, thought Beck, not for the first time, driving to Sidney. *The man is a millionaire. Maybe he would be interested in funding my research.*

It was a thought Beck said to nobody, nor did he allow himself to think about it much. It was an idea allowed to just fester in the regions near the top of his subconscious. Beck had another hope, a request that seemed more reasonable, one that would help him answer a critical question and one that had plagued him throughout his post-doctorate work : Was it possible to immunize a woman against becoming pregnant?

In other words, could you give a woman the proper antibodies to make it temporarily impossible for the embryo to implant itself?

The hope was that the vaccine could be used in Third World countries where the pill wasn't readily available but needed and wanted by impoverished women who didn't want to have fifteen children in the course of a lifetime.

Beck had another hope ...one that would help him answer a critical question ...: Was it possible to immunize a woman against becoming pregnant?

The Ohio State University had its own baboon colony, "and we thought that if we could get the antibodies, we could give them to the baboons," says Beck, from behind the desk of the company he runs, over thirty years later. "But to do that, we needed a large quantity of antibodies."

Beck needed to immunize cows to get the antibodies from the milk, which he would then give to the baboons at OSU. Ralph, meanwhile, not only had cows, he had some of the best milk producers in the world: Holsteins. At one point, Ralph had the largest herd of Holsteins in the state of Ohio, and as far was Beck was concerned, this was a man he should meet. For years, Ralph had been doing similar antibody work with immunization and bovines, developing what was always known as immune milk, in the name of curing arthritis, cancer, and other ills. Ralph's idea was based on the work of Dr. W. E. Peterson, whom he had first read about in 1958. There are some antibodies in cow's milk; if you could manipulate the cow's immune system, you could boost the antibodies in the milk, and make the milk even more healthy than it already is.

The way Beck saw it, Mr. Stolle might just be interested in helping with his work. Or he might have absolutely no interest at all, and the day might amount to nothing more than a long drive. Beck's confidence wavered throughout the trip.

But his confidence rebounded on meeting Ralph, who seemed surprisingly down-to-earth for a self-made millionaire running an industrial empire. He may have been 68 years old, but he looked maybe ten or twenty years younger to Beck, who now says: "I was impressed with how friendly he was. Not anything like I expected. We went from his office to a local motel where he kept a room and fully equipped kitchen. We had drinks, and he made lunch. We talked about his interest in immune milk, and I suggested some ideas on how he could advance the science."

What sealed their newfound friendship was Ralph's little personal touch of preparing and serving his own homemade German potato salad. It tasted exactly like the kind Beck used to have at his grandmother's house.

Ralph's charm was quickly working its magic on Beck, who now says, "This was the beginning of a long and life-changing experience for me. I have never met anyone else like Ralph. It is hard to explain. He made you feel relaxed and important."

What sealed their newfound friendship was Ralph's little personal touch of preparing and serving his own homemade German potato salad. It tasted exactly like the kind Beck used to have at his grandmother's house.

Each man received what he was hoping for. Beck kept offering counsel on the immunization; Ralph started work on inoculating his cows with antibodies that could be given to mating baboons. Meanwhile, Vern Stevens would work with the World Health Organization to develop a six-month vaccine for direct immunization of women against fertility, achieving the goal in 1978 and adding another link to Ralph's legacy.

And when Beck took his job at the University of Alabama, Ralph was only too happy to fund a baboon colony at the college. Meanwhile, Beck continued to offer his own advice on the research being conducted at Ralph's dairy farm. Which appeared to be going mostly nowhere. Yes, Ralph was producing a lot of milk that

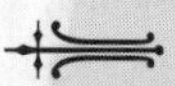

he claimed could help anybody's arthritis. The problem was, the scientific proof behind the claim was thin. "There was a lot of controversy over this," says Beck.

Indeed. One journalist, writing for *The Post & Times-Star* in 1966, began a column that opened: "Would you—if you were among the nation's fourteen million sufferers from arthritis—attempt to treat your disease by sitting in a "uranium tunnel" or by drinking "immune" milk from cows injected with certain vaccines, at $1.70 a quart?"

Although Ralph Stolle was never mentioned in the article, the reference to the immune milk was clearly aimed at him, and it didn't amuse at least one associate, who cut out the article, stapled it to the a memo pad sheet and sent it to Ralph, with one question scrawled on it: Shall we sue her?

There is no evidence that Ralph did. Maybe he appreciated freedom of the press and speech. Maybe he was aware the effort wasn't worth his trouble. And just as likely, he understood that the author had a point. There *were* a lot of fraudulent health claims out there. The uranium tunnel the columnist had mentioned referred to a fad from the 1950s, which was started by a Wisconsin resident with the unfortunate name of Kenneth Crook. He had a number of ailments and had read how people felt better after visiting mines that were mildly radioactive; and so, after visiting a mine repeatedly in Arizona and concluded that he, too, was healthier, Crook and his wife packed 2,100 pounds of ore in their car, drove home and built a uranium tunnel.

JUST ONE OF THE GUYS: RALPH (RIGHT) HANGING OUT WITH HIS FRIENDS AND CO-WORKERS AT THE STOLLE CORPORATION IN THE EARLY 1960S.

Crook charged $10 for eleven visits, and many people swore that it cured or eased pain stemming from everything from asthma to fungus. So many people believed in it that for a couple years, uranium tunnels began springing up across the country, and people began flocking to uranium mines. A Wisconsin tavern owner even spent $3,500 to add a uranium room to his bar. Even today, throughout the country, there are still uranium mines open for business, and some people insist there are health benefits to be found.

However, Ralph was no crook and didn't want to be thought of as a quack. But Beck was skeptical. "You don't have any science or patents backing any of this up?" he told Ralph over the telephone one day. "It's never going to be a business, unless you do more research and science." That made sense to Ralph, suddenly filled with purpose and prepared to try whatever Beck suggested.

Dr. Malik Sarwar was not so keen. It's impossible to know how Sarwar reacted when Ralph gave him the news, but because few people ever dared contradict or stand in the way of Ralph, it's easy to imagine that the scientist from Minnesota managed to crack an icy smile, and nodded in agreement with every word that his boss said. But he wouldn't remain silent for long.

Dr. Sarwar, now 61 years old, had spent over a decade running his lab without any interference, and perhaps complacency had set in, which would be easy to have happen when one is working in the idyllic rural Ohio countryside and shuttling over to the University of Kentucky as a research associate. Dr. Sarwar was coasting, and whatever fires had burned inside him at the start of his work were apparently now embers. Or maybe he simply felt it was unfair for this young upstart from Alabama to suddenly call the shots in a lab that he, Dr. Sarwar, had helped build from the ground up. Sarwar apparently said little to nothing to Ralph about his feelings, and at first, he went with the plan.

RALPH AND DOROTHY MAY HAVE NOT HAD A FOURTH CHILD, SIMPLY BECAUSE HE WOULD HAVE HAD TO EXTEND THE NAME OF HIS BELOVED FARM.

The plan was to immunize the cows against tooth decay. It hadn't been done yet, and Beck felt that they could get a patent, which they did, though they didn't turn their development into any sort of business. But after a few weeks of experimentation, Sarwar didn't show up to the lab one morning. When Ralph called to ask why, Sarwar reportedly explained it this way: "I'm not coming to work until you get rid of Dr. Beck."

What Ralph said to Sarwar isn't known, but Ralph didn't fire him on the spot. Instead, once he hung up the phone, he called Beck and explained what happened and asked if he would visit the lab and see if Dr. Sarwar had made any progress. Ralph flew Beck up from Alabama, and Beck did some checking and recognized that nothing had changed. The same antigens had been ordered as usual; the lab remained the same as it had for years. Ralph called Sarwar and asked him to come back and get with the program, or they would go on without him. Sarwar evidently said he would think about it, or he agreed to come aboard. But Sarwar never returned to the lab, and a week later, Ralph learned why. Sarwar had had a heart attack and died.

As Ralph explained this latest plot twist over the telephone, Beck took all of this in. And then came that second life-changing question, the one that sealed Beck's fate: Would you take over where Sarwar left off and work for me?

Ralph hadn't forgotten about the Stolle Corporation, and for the first four years in the 1970s he was hovering through the halls, overseeing workers, shaking hands, making deals, and whining and dining potential customers. One retired Stolle Corporation staffer, who doesn't want her name used, says that during those days, Ralph didn't want coffee or water at an employee's desk. He believed it interfered with a worker's concentration. "And so whenever we knew he was coming, we hid the coffee pot," the worker said, chuckling at the memory. Meanwhile, Janet Aikin, who came to the Stolle Corporation in 1971, remembers the scent of filet mignon roast, one of Ralph's specialties, wafting through the hallways, intended for a bigwig client who was on the way. Even though it was the lunch hour and everybody was free to go, "nobody wanted to leave the office," says Aikin.

[Ralph] had no intention of sitting out the rest of his years in a recliner and watching daily episodes of The Price Is Right . . .

But Ralph wanted to leave, or he felt that it was time to exit. By 1974, Ralph was 70 years old. He had no intention of sitting out the rest of his years in a recliner and watching daily episodes of *The Price is Right*, and yet he knew he couldn't run the Stolle Corporation forever. And maybe he didn't want to. Dorothy had never enjoyed being alone, and by selling the company he wouldn't have to make the trip to Sidney every day. Whatever Ralph's exact reasons were, he felt it was time to step back, and so he sold the remainder of his stock of the Stolle Corporation to Alcoa, a company that he had had strong ties to since the 1930s.

It was a monumental decision, one that weighed heavily on Ralph because it meant he would no longer have control over the fates of his extended family. "When Ralph was running it, you were never laid-off. You always had a job," Don Boyer would say, years later, after his retirement. "I remember one woman who

was in her 70s, and she was going to be forced into retirement, and Ralph stepped in and said she could work there for the rest of her life." Once Ralph was gone, that wouldn't be happening. And he knew it. Once he left, the Stolle Corporation would lose its guardian. "Ralph took care of his employees like a father figure, even if a problem that came up was the employee's fault," observes Ralph's pal, George Henkle. And the Stolle employees were about to watch their patriarch retire. Nobody really liked the idea, and it's easy to see why. As Harold Hall, who worked for Ralph for eighteen years, observes, "I never witnessed a display of anger or shock. He was even-tempered at all times."

But nobody could begrudge a septuagenarian who wanted to step back a little, or who wanted to pursue other interests in his golden years.

He would prove to the world that milk could immunized to fight disease, particularly the most deadly of them all. Cancer.

Which is exactly what Ralph wanted. He had anodized aluminum and turned it into gold. He had improved everything from automobiles to refrigerators to fork-lifts. He revolutionized the soda and beer industry. But there was another beverage that he saw more promise in, a liquid that appeared to have as much going for it as the legendary Fountain of Youth.

Milk.

But not just any milk, of course. Ralph believed his white stuff could not just alleviate arthritis, or improve the immune system. He didn't just think it was possible to develop milk to make one a more healthy person. He was after bigger game. He wanted to achieve something that would usurp everything he had done previously. He would gather and collect all the data Beck needed. He would prove to the world that milk could immunized to fight disease, particularly the most deadly of them all. Cancer.

Beck took the job. Once Sarwar was gone, Ralph had nobody else close to him that he could turn to. "I was kind of locked in," says Beck, who later that year moved his family to Cincinnati and started managing the lab at the dairy farm while trying to continue the research he had begun at the University of Alabama. A year later, Beck hired an immunologist, who still is working for him, and the research program was truly up and running. It was then that Ralph initiated a conversation with Dr. Beck.

"OK, I want to make this a business," said Ralph. "I want this to go on after I'm gone."

The way things had been going, the farm was nothing close to a business. It was the one area of Ralph's life that had yet to make any serious money, but that was only because he hadn't applied himself. From 1941 until 1968, as Ralph would later write, San-Mar-Gale Farms hadn't made a profit. This was troubling because in 1963, the friendly neighborhood Internal Revenue Service came calling. They informed Ralph that his farm had to make a profit in at least one year in the near five years, or it would be declared "not a business venture, but a hobby." All the previous farm losses would come back to haunt the Ralph J. Stolle Company. It was chilling news, and Ralph set out to make a profit. Which he did. In the last year. It took most of the time given to them by the IRS, but the farm bred a bull named Capsule, which they believed would improve the breed in milk production.

"In early 1968, we solicited the breeding associations to purchase Capsule," wrote Ralph. "Several associations were anxious to purchase him but were shocked at the price we were asking. They pointed out that the highest price ever paid for a bull up to this time was $50,000. To show a profit for that last year, we had to get $250,000 for the bull."

Ralph ultimately sold the bull to Carnation Farms, the company that makes milk and baby products. He was given $250,000 for it, Carnation never indicated that they were displeased with the sale, and the IRS dropped its charges. Once more, Ralph had escaped doom. But with Beck on board, Ralph was ready to get serious.

The two hammered away at details. Ralph still wanted to give his powdered immune milk away to friends, family, or anybody who truly felt they needed it. That was important to him. He also wanted the milk developed as a drug.

RALPH AND GOVERNOR JAMES RHODES HAD A SOLID FRIENDSHIP, BEFORE AND AWAY FROM THE CAMERAS.

Beck agreed to the unusual business model of giving away the product, for the indefinite future. But he didn't think the drug idea would work. Well aware of the complex Federal Drug Administration maze they would have to navigate to ever have hope of developing a drug on a dairy farm, Beck felt they'd have better luck marketing the milk as a nutritional supplement. Ultimately, that's what would happen. Meanwhile, Ralph continued to offer free milk to anybody who asked for it, but he requested that the

person report in every month on how they were feeling and if the milk was helping. That way, they could have case studies and anecdotal evidence to help support the scientific testing being conducted at the lab.

In part, what drove Ralph to introduce his milk to the world wasn't a desire to get rich—he already was—but simply because he had a lot of misgivings about the medical community, says Beck. "He didn't have a high opinion of doctors. He always liked the idea that he could come up with something better than the doctors." One can't help but think Ralph's opinions were probably formed young. Doctors hadn't been able to stop tuberculosis from staging a coup in his body for an entire year. They hadn't been able to save Dorothy's two brothers from illness. Ralph approached medicine with much of the same principals he had in business. f there was a problem, fix it.

That was his thinking with the problems of youth. Kids who became involved with crime or who were slacking off in school and life needed to get off the streets. They needed a wholesome environment to grow up in. Clearly thinking of his days at the Hi-Y Club and participating in his church's activities, and probably not up to speed on what adolescents of the 1960s and 1970s were up to, Ralph told Henkle: "These kids need some place to go rather than saloons or dance halls."

And so when Ralph wasn't trying to advance the cause of his milk, he was helping, in his adopted hometown of Lebanon, to plan a place for children to spend time after school. But this wasn't to be just any ordinary, run-of-the-mill children's club. That wasn't Ralph's style, especially at this stage in his life. He had seen too much, and done too much, to give Lebanon just any old meeting place. And so Ralph teamed up with his buddy, George Henkle, and several other local community leaders, Ellis Thompson, Charles Hamilton, Marvin Young, Mike Norris and the most prominent Lebanon citizen, Neil Armstrong—yes, *that* Neil Armstrong.

ONE OF RALPH'S PROUDEST ACCOMPLISHMENTS—THE COUNTRYSIDE YMCA IN LEBANON, OHIO, LATE 1970S. IT WAS ONE OF THE LARGEST Y'S IN THE NATION, POSSIBLY THE LARGEST: A 126-ACRE SITE WITH AN 82,000-SQUARE FOOT BUILDING. IT CAME COMPLETE WITH A FISHING HOLE.

Ralph and Armstrong had met in 1972 through a mutual friend, a realtor named Ellis Thompson. Armstrong had moved to the area after accepting a job at the University of Cincinnati as a professor of aerospace engineering, and

he and Ralph hit it off. They may have not become best friends—indeed, Ralph was twenty-six years Armstrong's senior—but they did become close. Armstrong puts it this way: "Ralph—R.J.—was probably a seven on a ten scale."

They played golf together every once in awhile, and they hunted pheasants on Ralph's farm, and Armstrong has fond memories of Ralph's cooking. Once, Armstrong invited Ralph to one of the moon launches. Slim Buchanan recalls that Armstrong also visited Ralph in Florida a few times; Ralph had come to enjoy the warm and sunny climate and frequented the Sunshine State whenever he was able. What struck Slim about Armstrong's visits to see Ralph was that "they never talked. They never said nothing. They'd just sit, and maybe read some newspapers, for a few hours."

. . .probably not up to speed on what adolescents of the 1960s and 1970s were up to, Ralph told Henkle: "These kids need some place to go rather than saloons or dance halls."

But for a time, there was plenty to talk about. For starters, in what shape and form would this envisioned children's Utopia take place. Recalls Armstrong: "It was not clear in the beginning under which organization umbrella the facility would operate. After studying the alternatives, we concluded the YMCA was the most appropriate organization. We then visited a number of Y's around the country—usually using Ralph's airplane—to learn the best and worst features of their building configuration and governance structure."

That Ralph had such a famous friend surely elevated Ralph in other people's eyes, but it isn't that surprising. A famous retired astronaut has to retire *somewhere*, and Armstrong settled into Lebanon, Ohio, with the ease that most people reserve for a comfortable chair. After awhile, he just became part of the town. And as it was, Ralph had a knack for befriending just about anybody, whether it was a farm hand or James Rhodes, one of Ralph's best friends and who was governor of Ohio from 1963 to 1971 and then from 1975 to 1983. Rhodes and Ralph were both Republicans, and ardent followers of the cause. Ralph gave money to both local parties, but he always donated to national causes. Ralph was a great admirer of Ronald Reagan, and he approved whole-heartedly of George Bush and would likely have appreciated his son, Bob Mays theorizes. Books about the Kennedys were in his bedroom, although Ralph was probably too busy to do anything but glance through them. One story, that says more about Ralph's patriotism than his feelings towards a Democrat president, involves George Henkle, who remembers heading to an Ohio State University game when they learned on the radio that President Kennedy was shot. They were passengers in one of those buses lent out by the Secretary of Transportation, and Ralph instructed the driver to take them to Sidney

instead; everybody filed into the Wagner House, where Ralph's suites were, and they spent the weekend reflecting on the nation's ugly turn of events and still managed to have a good time, but in a more low-key way than had been planned.

That Ralph and the governor were friends is indisputable, but business was often worked into the equation. Ralph had land in Marysville, Ohio, and when it came time for him to sell it, Rhodes found him possibly the best buyer in the world: Japan. Ralph sold his land to Honda, which still has a plant there today. (The Japanese businessmen were very impressed with Ralph, recalls Dick Pope. They thought that anybody who had as much land as Ralph did must be something close to royalty, and whenever they did business, they kept bowing to him.) Ralph was able to use a lot of his political connections, whether it was for friendship (football games), business (his connection to Honda) or establishing a YMCA for a community in need.

For when it came time to open the doors of the YMCA on September 15, 1978, James Rhodes naturally made an appearance. It was a quick one; in a short speech, he accurately called the place "unique, exciting, bold, and imaginative," he cut a ribbon, and then he was off and away, campaigning elsewhere. But even without Ralph's involvement, it was a smart political decision for Rhodes—the Countryside YMCA was one of the largest in the nation, possibly the largest: a 126-acre site with an 82,000-square-foot building. There was a reason it was called the Countryside YMCA—it came complete with a fishing hole and another lake for sailing and canoeing and an outdoor Olympic-sized pool for swimming—and that could accommodate up to 3,000 people at a time. There was a small orchard for farming and garden projects, and another small forest for family picnics. True to Ralph's ideals, there was also a small dairy farm on the land. It had already been on the property when Ralph bought it up, but there was no way he was going to tear it down. Inside was another large swimming pool, eight handball courts, a running track, weight-lifting equipment areas for both men and women, a large gymnasium, and the list went on.

"Howard used to say, 'Ralph's going to kill himself, working so hard.' Well, Howard died first."

It was a gargantuan undertaking to construct the facility between the time Ralph and George started discussing the creation of a new YMCA in 1971 to the time the doors actually opened. Later, during a two-year planning stage, every Sunday night, from eight o'clock to ten and sometimes later, Ralph, George Henkle, Neil Armstrong and other community leaders met at the various men's homes, although usually at Ralph's farm, according to Armstrong, and they planned out the YMCA "in painstaking detail," says Henkle.

Ralph and his colleagues visited twenty-seven YMCAs in all, to determine what they wanted at their facility, as well as what they didn't. Blueprints were drawn up, disbanded, drawn up, disbanded, and drawn up again. Mission statements were hammered out. It would be debt free when it opened. There would be scholarships available to the needy. There would be an endowment fund. But most of all, the emphasis would always be on advancing the cause of family unity.

Henkle's primary duties were in fundraising, and Ralph's focus was on construction, since he had built more than his share of office buildings, warehouses, and plants in his time. Armstrong also assisted mightily with the construction phase and in "general organization," which all did, says Henkle.

THE STOLLE SIBLINGS: TOP FROM LEFT TO RIGHT: LES STOLLE, HOWARD STOLLE, RALPH. BOTTOM, LEFT TO RIGHT: IRMA PENDERY, CHARLOTTE PENDERY, MARGARET HILLIKER.

And throughout the process Ralph kept leading the charge, insisting on a YMCA that would be among the best in the nation, says Henkle, who is full of praise for his old friend: "I once told Ralph, 'You're right ninety-six percent of the time because you've had so much experience. That's why we all fall behind you—as a leader, you know what you're talking about.' And after observing Ralph, as well as Neil Armstrong, I decided that the mark of a genius is somebody who applies lessons he's learned from experience to the issues at hand. But Ralph was also persistent, like Edison, who had 799 failures before the light bulb. Ralph would never give up on anything he would do—and that's another secret of a genius. Like Churchill, Ralph never, ever gave up. And that's a good lesson for all of us to learn."

Ralph was 74 years old when the YMCA finally welcomed its guests in 1978, but he had hardly considered that the capper to his career. The same year, he started branching out in the nearby city of Wilmington, creating a new business, Hydroelectric Lift Trucks. He also became a trustee at Wilmington College. And he hadn't given up on his immune milk. Steering Beck's science into a business, Ralph incorporated Stolle Research and Development.

Ralph's closest sibling would shake his head at all of this activity, recalls Henkle with a chuckle. "Howard used to say, 'Ralph's going to kill himself, working so hard.' Well, Howard died first."

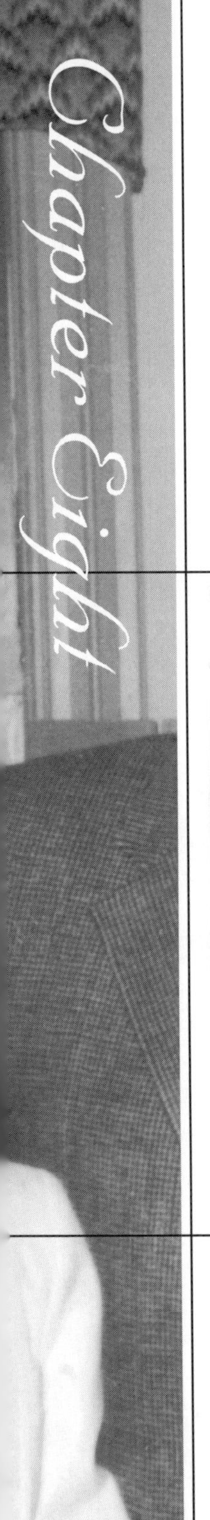

1980s & MORE

"To live life to the end is not a childish task."

— BORIS PASTERNAK, *DOCTOR ZHIVAGO*

RALPH, AND HIS SECOND WIFE, PAULINE.

IN THE YEARS AFTER RALPH DIED, for any one who knew him, a drive around Sidney—the city he used to practically own—was a melancholy and bittersweet trip. At 109 Highland, what used to be Sidney Tool and Die and part of the Stolle empire, is just a shell of a building. Farther down on Highland Avenue is a plant that's been closed for almost a decade, a forgotten brick building with boards on the windows. A sign reads, *The Stolle Corporation*, but more prominent are those that read, *Keep Out* and *No Trespassing*. Overlooking the building is a water tower with peeling paint.

It isn't like this everywhere. In many sections of the city, Ralph still lives. The headquarters of the Stolle Corporation is still at 1501 Michigan Street where Ralph quietly encouraged everybody to be better than their best, and where Kaminski worked in his office, designing diagrams for everything from the easy open pull tabs to developing a complete aluminum siding production facility.

Stolle Avenue is in a nondescript part of the city, yet it's located in a very pleasant place in town, an area that, by the looks of things,

will probably see a lot of growth in its future. There are well-manicured green fields nearby, with nothing on them but grass, land the Stolle estate still owns. And nearby, at the Christian Academy High School, is a gymnasium named after Ralph because of all the money he poured into the institution over the years. Until shortly before his death, Ralph was giving money away to any place that he felt needed it.

Ralph's beloved Dorothy died June 27, 1983. As Lois Henkle observes, "Ralph never went anywhere alone," and now his lifelong partner since the 1920s was gone. San-Mar-Gale Farms was now a very quiet, very empty place. Dorothy had died of lung cancer, and while she had been in poor health for some time and hospitalized for about six weeks from the affects of cancer, her death still felt sudden and unexpected. "Daddy really took her death very hard," recalls Mary Jo. "He couldn't be himself. Grandchildren would take turns, and Brad [a grandson] lived with him for quite awhile."

For about a six-month stretch, Lee Beck was having dinner every few nights with his old friend, who was a very lonely man. Beck got to know his friend and business partner extremely well during this period, and it was then that he came to see Ralph as he was, a super human but not super-human.

[During this period] Beck . . .came to see Ralph as he was, a super human but not super-human.

"Ralph had an inferiority complex," says Beck, a statement that will be surprising to anybody who knew him. Ralph was rich beyond belief, successful beyond measure and well-liked—you get the idea that if he had lost everything, and the bank examiner was threatening to throw him in the clink, it wouldn't have been difficult to round up a couple hundred of his closest friends and family to bail him out, just the way it happened in *It's a Wonderful Life*. But, yes, Beck insists, Ralph had an inferiority complex. "He had it around people who had more education than he did," says Beck, and he observes that many of the entrepreneur's friends were self-educated, much in the way that Ralph was. In fact, at one point late in Ralph's life, Miami University in Oxford asked Ralph to come on the board of trustees, but he refused because he was embarrassed that he wasn't a college graduate and thus felt he wouldn't have anything to offer.

Ralph was outliving everybody. His comrade-in-arms at Alcoa, Fritz Close, passed away in 1986. Ermal Fraze, who received most of the credit for the easy-open pull-tab, passed away in 1989. His lifelong friend, Reuben, died on

December 8, 1993. After awhile, it was clear that everybody who had been with Ralph in the beginning, and even in the middle, would no longer be with him at the end.

It must have been a depressing notion, and Dorothy's death was a punch in the stomach, but it didn't get Ralph down for long. He married again, not very long after Dorothy's death, a lovely, elderly woman named Pauline Yergin. But she, too, died, in 1990. In that case, friends say, Ralph literally wore her out with his active social life. Even well into his 80s, Ralph was running the Ralph J. Stolle Company at his farm and checking in on the facilities of the Ralph J. Stolle Corporation that Alcoa owned in Sidney. He was traveling to Florida when he wasn't back in Ohio, advancing Stolle Milk Biologics with Beck and transforming it into an actual business. There were trips to New Zealand, where the immunized milk was being made, and Thailand, where the immunized milk was being enthusiastically drunk. Pauline didn't blaze every trail that Ralph did, but the quiet retirement that she expected to have with her new husband was nowhere to be found. There were always parties that Ralph was throwing, and there were always people at the house. "She didn't know what she was getting into," theorizes Lois Henkle.

Ralph, a widower again at 86, was attracted to Kathleen Porter, a county court judge. Since she was 47, he probably figured she had a shot at keeping up with him . . .

Which might be in part why Ralph, a widower again at 86, was attracted to Kathleen Porter, a county court judge. Since she was 47, he probably figured she had a shot at keeping up with him, and so he convinced a mutual friend to fix them up. They married in 1991. "He was so alive and so vital," recalls Kathleen. "He didn't seem his age at all. And then my late husband was thirteen or fourteen years older than me, and my father was twenty years older than my mom, so to fall in love with an older man was nothing strange to me." And like so many other people before her, Kathleen found Ralph "captivating."

As for Ralph, he was equally captivated. Typically, Ralph wouldn't approve of an older man marrying such a younger woman, and his third wife yet, says his friend Dr. Lee Beck. "He was the kind who would frown on that. But when it's you in that situation, it's different."

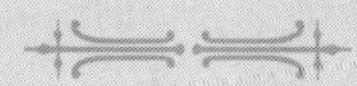

In Lebanon, Ohio, what was once called the Countryside YMCA is now the Ralph J. Stolle Countryside YMCA. It was renamed after his death; while Stolle was still alive, he wouldn't allow the facility to carry his name. He was a man, who in general, did his best to stay humble, dressing simply and refusing to buy into status symbols, like owning a Cadillac (he believed the only type of journey one could take in a Cadillac was an ego trip). Once in Ralph's later years, a journalist declared his intention to write a book about the multi-millionaire businessman; Ralph apparently was too kind to refuse him, but he refused to talk about himself. For two days, the writer followed Ralph, who pretty much ignored the poor fellow, until he finally got the hint, gave up and went away.

FROM LEFT TO RIGHT: CONGRESSMAN ROB PORTMAN, RALPH, KATHLEEN PORTER STOLLE AND PRESIDENT GEORGE H. W. BUSH. RALPH WAS A LIFELONG SUPPORTER OF THE REPUBLICAN PARTY.

And so it was within Ralph's character that he would shy away from having a building named after him. And what a building. It's one of the biggest YMCAs in the world, a facility that now covers over 200,000 square feet. (For some perspective, when the Concorde flew, it needed 11,790 feet of runway to take off.) There is an indoor soccer facility, indoor swimming pools and three outdoor pools, one of which has a playground in the middle of the water. The activities that take place here include everything from foreign language instruction to CPR, soccer, scouting, aerobics, and arts and crafts. The Utopia for families, the place for kids to get out of the saloons and dance halls, and in more recent decades, away from drugs and drinking, is just what Ralph hoped it would be.

In later years, Ralph started to buckle and accept the idea that he had something to offer college students. He bought land in and around Wilmington, Ohio, and so he started paying attention to its college. That was in part, because one of his many business associates worked with the college, and he started harping on Ralph to get involved. In the late 1970s, Ralph became a college trustee, and that's when Campbell Graf met him. Graf, who joined the university to start a cooperate education program, was inspired by all that Ralph had done, and how nothing seemed to faze the guy.

"He maintained a certain calm and poise that was just remarkable," says Graf, who remembers hearing a story of an ice storm doing major damage to a plant that the Deerfield Manufacturing Company was building (yet another business that Ralph picked up along the way). "The same morning, some personal thing went wrong—I'm not sure what—and Ralph just says, 'Things like that just happen.' He was always calm and cool and simply didn't get rattled."

SURROUNDING RALPH ARE HIS DAUGHTERS, FROM LEFT TO RIGHT: GAIL NORRIS, MARY JO CROPPER, SANDY PERRY

Graf goes on: "I always saw Ralph as a latter-day Benjamin Franklin. He was an inventor, an entrepreneur, and a very persuasive salesman, interested in anything and everything." And, coincidentally, both share the same birthday.

Over the years, Ralph donated generously to Wilmington College, but probably never more so than when he gave them $1 million in his will, after he died. The money went towards what the college calls the Ralph J. Stolle Free Enterprise Program, which aims to work with local entrepreneurs in developing business plans, conducting marketing analyses, crafting strategy, and such. It's taken a little while for the program to take off, but it appears to be growing stronger and not weaker. But the gift was just one of many that Ralph gave. He generously gave to the Cincinnati Children's Hospital, so much that his name is still on a plaque of donors. Only Procter & Gamble appears to have given more than Ralph did. Nobody knows how much money Ralph gave away during his lifetime, but it was clearly millions of dollars. He funneled money to the YMCAs in Sidney, Wilmington and Fort Thomas. He gave to churches. And he gave to people who simply needed it. He was a millionaire with a ridiculous amount of money, but arguably, he wasn't ridiculous in how he spent it.

And sometimes when he gave away money, it turned into a profit for him, bouncing back like a money boomerang. In the last year of Ralph's life, he conducted an interview with two ladies who were from Wilmington College and planned to write a piece about him in a newsletter or collegiate magazine. In any case, Ralph shared a story of how one company, the Sheffer Corporation, came about to be something that he owned. A friend who also ran a business with a partner, had approached, said Ralph, "and they were building cylinders in their basement, and they needed money, and so he came to me and wanted to know if I could lend them money—I think it was $10,000. Well, they went through that. And then it was another ten, and another ten, and finally I said, 'Look, I want to buy half interest in the company, and if you ever want to sell, I want the first refusal to buy.' And so later they wanted to sell, and that's how I bought that company."

Granted, Ralph wasn't a complete softie—he recognized the worth of the business to his own. His friend was making cylinders, which Ralph knew in the back of his mind that he might someday use for his own company. But it's just one of many examples of Ralph trying to help somebody else make a buck and, in the process, making one for himself.

During the interview, which was long on chatter but short on substance, Ralph was asked about his creativity, and he ended up explaining the secret of his success: "My mind usually ran along innovation," Ralph told them. "We built a company from one man, myself, to three or four thousand people when I sold it, and it was all built on innovation, things we'd think about and put into practice."

Indeed. Ralph was on hand in 1985, at a Stolle Corporation ribbon cutting ceremony, which was attended by Ohio Governor Richard Celeste; the Stolle Corporation had opened up a $35 million Memory Products Division, which manufactured memory discs used in computers—made from, of course, aluminum. About the same time, *The Cincinnati Enquirer* did a story on the Stolle Research Corporation, which had for some time been experimenting with animal embryos.

He was a millionaire with a ridiculous amount of money, but arguably, he wasn't ridiculous in how he spent it.

These were vast experiments, being done not only at the farm but at the University of Alabama with the baboon colony Ralph had funded. In 1983, two mother baboons made history at the University of Alabama by having one animal conceive an embryo and the other carrying it to term—after having it transplanted into her uterus. (That same year, Ralph was having the laboratory built where

Stolle Milk Biologics now stands. Beck and a colleague, Danny Lewis, came to pick Ralph up so they could go look at the site. Ralph said, "I'm not sure I can stand up." Beck asked if Ralph was having a back problem. "No," Ralph said. "I just got the bill on the stainless-steel water system for the new lab." *Rim-shot*. Ralph may have missed his calling as a comedian.)

Ralph's life, and life with Ralph, had become one long, glorious party that seemed like it would never end.

In 1985, the *Enquirer* was reporting on an injectable microscopic capsule, which would inhibit pregnancy and eliminate the need for a daily pill. Still unrealized, but it's a good example of how Ralph's interests seemed to have no bounds.

He never stopped innovating. Even when Ralph turned 90, he was running his business full tilt. Kathleen has vivid memories of traveling with Ralph to Taiwan and making another journey to Hong Kong and Beijing, all in the name of furthering Stolle Milk Biologics. But they had fun overseas, too, says Kathleen, especially when trying out restaurants. "You never knew what you were eating," laughs Kathleen, "and Ralph would eat anything."

And Ralph would cook anything. Even into his 90s, he was preparing major meals for parties that appeared to have been designed for half the planet. Kathleen especially recalls an appetizer that her husband used to make. The recipe went something like this:

One can of chili sauce. (Or twenty cans. It really depends how many people are going to show up.)
One packet of crab meat. (Or twenty.)
Philadelphia cream cheese.
Mayonnaise
Mix up the chili sauce, the cream cheese and the mayonnaise .
Put in a small dish, and put the crab meat and chili on top.
Serve.

Ralph had other unusual fare that he enjoyed fixing. There was what he called Oyster Stolle: "Take a raw oyster and dip it in cracker crumbs, and put an anchovy on top it," Ralph instructed the two ladies interviewing him. "Then put it in the oven with some cheese."

"It's really outstanding," Dick Pope assured the ladies.

Ralph also loved making meatloaf, mashed potatoes, and biscuits. The meatloaf, remembers Kathleen, would always have something new in it, and always vegetables—sometimes carrots, onions, even okra. He also devoured shrimp, cooking and munching on them. And Neil Armstrong fondly remembers "eating his legendary hasenpfeffer," a dish, of course, utilizing rabbit. "Fry it in the fryer," Ralph told the ladies, "and then put it in gravy. A lot of people love it, though some people won't touch it."

IT WASN'T ALWAYS JUST ABOUT POWDERED MILK. RALPH IS FEEDING ICE CREAM TO HIS GREAT-GRANDDAUGTHER, JULIA.

Every month, for about a week, Ralph took Kathleen to Florida, and every Fourth of July, he threw a family reunion bigger than the last, and most weeks, Ralph and Kathleen took long Sunday drives after church. They dined out several times a week and took long walks on the farm, studying the cattle and angling for catfish and then frying it up that night. And, of course, Ralph doted on his grandchildren and then his great-grandchildren. Ralph's life, and life with Ralph, had become one long, glorious party that seemed like it would never end.

1994-96 EPI

"I never did a day's work in my life, it was all fun."

— THOMAS EDISON

EVEN INTO HIS 90S—AND, YES, THIS WAS TAKEN EARLIER—RALPH NEVER STOPPED WORKING.

THE YEAR OF 1994 was winding down when Kathleen convinced Ralph to see a doctor. He had been suffering a sharp pain in his left arm, and they had tried every remedy that they could think of, but nothing seemed to work. And so Kathleen finally said: "This is it. I'm calling the doctor. Because I trust doctors!"

Up to the end, Ralph had his suspicions about general practitioners. In fact, the following year Ralph would comment on a visit to a Florida hospital to those two ladies from Wilmington: "I feel 100 percent better than I did the last time you were here. I had a lot of aches and pains, and being in the hospital in Florida was a real mistake. They just constantly come in and work on you. And I'd get upset all the time."

But even though Ralph held little regard for doctors, he humored his wife and went, and he couldn't have liked them any more after his check-up. What news he was given was bad. There was, the doctor said, a tumor on the back of Ralph's neck.

Fortunately, Ralph was given radiation treatment, and the tumor responded, and it was quickly gone. But just as quickly, a month later, right around Christmas, Ralph learned that he was still sick. He had prostate cancer.

It was a cruel blow; after spending almost half his life drinking his milk and trying to discover a cure for cancer, he had ended up getting it himself. But he said little, if anything, to anybody. "Cancer," says Kathleen, deep in thought. "I don't think that word was ever used. I don't think we ever actually said that word."

She shakes her head, full of regret. "If he had gone to the doctor sooner, they might have done something. But by the time he went, it was too late."

The new year arrived, and Ralph gamely went off to New York City's Sloane-Kettering Hospital every two months for treatments. Exactly what was done, Kathleen isn't sure, though she knows he didn't receive chemotherapy, radiation therapy, or an operation, which makes one wonder what was left. But Ralph was insistent that he didn't want his wife to come along on these trips. Fortunately, for a time, it seemed that whatever the doctors were doing was working, for he had rebounded. He had aged, certainly, but the mischievous boy who once stole pies from Daisy Schmidt's window and rearranged the furniture in his minister's house was still somewhere in there.

. . .after spending almost half his life drinking his milk and trying to discover a cure for cancer, he had ended up getting it himself.

It was May of 1995 when Kathleen, overseeing court cases on the bench, received a telephone call that Ralph had become sick. "Something happened... I'm not even sure now. It wasn't that bad," says Kathleen, who even now sounds a little shaken. Kathleen managed to get a judge to take over for her while she accompanied Ralph to the hospital. Bad or not, Ralph was 91 and still frail from radiation treatments. He refused to think about it, or to at least let anyone know he was thinking about it, but the end was getting near.

Kathleen bought her husband a scooter, and Ralph traveled on that around the farm and to his office, so he could conduct meetings with his long-time assistant Rita Hauck, Lee Beck, or whoever was on the roster for that day. "He always had his finger on the pulse of everything," recalls Kathleen.

It was September when during another stay at a hospital, Ralph slipped and broke his arm. "He had a lot of pain from that, and he didn't want to move," says Kathleen. "The last few months were difficult. He required quite a bit of care." And so, for the first time in Ralph's life, he slowed down. Kathleen had the door to Ralph's bedroom enlarged so that his bed could be wheeled up to the living room's windows, and Ralph could gaze out on a pastoral wonderland, where he watched his geese enjoy the pond and his herd of Holsteins grazing in the fields.

Ralph died in his sleep, in his beloved ranch home, the following year, January 13, 1996. He was buried on his birthday. At the funeral, Reverend Bill Cain spoke eloquently about Ralph, summing up his life in a mere sentence: "Idle dreams are for idle people, and Ralph Stolle was anything but an idle individual."

As the centennial of Ralph's birthday approached, in the well-to-do community of Blue Ash, Stolle Milk Biologics would continue to thrive. The business was sold by the Stolle family shortly after Ralph's death. Nobody quite seemed to know what to make of it, which is why the selling of the company may have not only been a sound financial decision, but a humanitarian one as well. Stolle Skim Milk Powder is on sale throughout pockets of the world; children and adults in Taiwan, Malaysia, Korea, and Japan are benefiting from the powder as well as from a product called Stolle Whey Protein Isolate-Plus. In America, Ralph's immune milk has been tangled up in the matrix of the Federal Drug Administration, except for a Stolle Milk Protein Concentrate, which is sold in the United States, as well as the aforementioned countries. And in recent years, Stolle Milk Biologics is looking into putting its milk protein concentrate into products such as yogurt, soymilk, even applesauce.

"Idle dreams are for idle people, and Ralph Stolle was anything but an idle individual."

Stolle Milk Biologics was sold to the Spencer Trask Specialty Group, LLC. The mission of Spencer Trask isn't all that different from Ralph's during the 20th century: "To discover and invest in the ideas that will shape the 21st century." Ralph would undoubtedly approve of Trask overseeing his work. In 1879, Trask financed Thomas Edison's discovery of the light bulb, and in later years, Ralph found himself greatly admiring the inventor. And while another company owns Stolle Milk Biologics, Ralph hasn't been forgotten. His portrait hangs on the wall in the building, greeting visitors as they walk in the front door.

Dr. Lee Beck still runs Stolle Milk Biologics, and he clearly still thinks fondly of the man who sent his career in a wildly different direction than it had been heading. "Ralph was always interested in looking for an alternative approach," says Beck, when describing the secret to his friend's success. "Of all the things he did, Ralph felt that people would remember the Stolle milk the most," says Beck. "And it's being sold everywhere in Asia. And I think he was right about that. That's what people are going to remember him for."

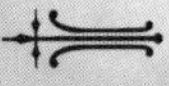

Ralph's friends and family remember him for other reasons, and in their own special ways. One year after Ralph's death, Dick Pope, Elton Kaminski, and a couple of other buddies of Ralph's traveled north to Canada and went fishing. "We went to the same place on the shore where we always had lunch, and we built a fire," says Pope, who had been going fishing with these friends—and Ralph—for twenty years. "And we all did a toast to Ralph. It was one of his favorite places."

Dr. Lee Beck never goes long without thinking of Ralph in some way. Not only does he work at the company with Ralph's name attached to it, he sees Ralph's portrait every workday. "I live on Ralph's farm," he says. "He gave me the land to build my house." Dr. Beck thinks a lot about Margie, too, and the times they both spent with Ralph before he got sick. Margie passed away just a couple of years ago.

ONE YEAR AFTER RALPH'S DEATH, HIS FISHING BUDDIES GOT TOGETHER TO REMEMBER HIM.

Mary Jo Cropper can't ever watch a boxing match without seeing her dad throwing imaginary punches, and she remembers her father whenever she sits in Ohio Stadium, watching the Bucks. "My dad is right there," she says. "He made every effort to get to as many games as he could, even when his health was failing." And, of course, she thinks of him when she is at the farm, and always on his birthday. Mary Jo's daughter, Amy, remembers visiting her grandfather in Florida and drinking the immune milk for breakfast and sneaking chocolate into it. Her brother Spencer remembers being 10 years old, working at San-Mar-Gale for some extra money and having Ralph spirit him off one day in his plane, taking the boy on a business trip to Myrtle Beach—and still paying him for a day's work.

"I remember him every day," says Kathleen, adding, "I have several pictures on my dressing table that I look at each morning… I am reminded what a wonderful husband he was and what effect he has had upon me."

Ralph's family and friends remember their patriarch and pal for taking them places they never would have gone otherwise, for urging them to work hard but not forgetting to have fun and do good things for other people, for communing and cavorting with buffalo and baboons, and mostly, for showing them what a person could do, if he or she had the mind and will.

HARD ACT TO FOLLOW: ALMOST 10 YEARS LATER, KATHLEEN PORTER STOLLE HASN'T REMARRIED. "I REMEMBER HIM EVERY DAY," SHE SAYS.

Beck is absolutely correct that Ralph was always looking for an alternative approach. Ralph Stolle became the prince of aluminum in a kingdom ruled by steel. He could have tried to improve the already-existing machinery that mass-produced the pull tab, but he entrusted his engineer to take the risk of dreaming up something altogether different. He could have decided that children in his town were fine without their own YMCA, or that throughout Northern Kentucky and Ohio, churches, schools, universities, and hospitals didn't really need his help. And he could have figured that as he settled into the life of a senior citizen, he had done enough and not bothered with trying to improve the rest of the world. Instead of researching cows and antibodies, Ralph could have driven to the store like everybody else, bought a two-gallon jug of milk and called it a day. But he didn't.

It's as if he were aware that he couldn't live forever, and so he decided to live for everybody.

It's as if he were aware that he couldn't live forever, and so he decided to live for everybody.

Prologue

FOOTNOTES

*Electroplating is often spelled as electro plating and electro-plating, as electroplating enthusiasts will surely note and possibly shake their head at my spelling. But I'm going with my preference, as well as the Merriam Webster Dictionary's version.

ENDNOTES

1 Tie, white shirt: Interview with Rita Hauck in April 2003.

2 Seven million cans: Letter from Elton J. Kamanski, written in mid-October, 2003.

Chapter One

ENDNOTES

1 Description of Ralph: Interview with Tom Anderson, on September 26, 2003.

2 Uncomfortable in crowds: Interview with numerous friends and family who said pretty much the same thing.

3 First mention of Kunigunde: Kunigunde lived to see her name spelled a dozen different ways. Her name is spelled this way on Ralph Stolle's birth certificate, and I'd like to think that they got it right.

4 www.familysearch.org, the web site run by the Church of Jesus Christ of Latter-day Saints, one of the best genealogy sites out there.

5 Ralph Stolle's birth certificate.

6 Johann: Ibid.

7 Kunigunde's birthday: www.familysearch.org

8 Frank Stolle's birthday: A fantastic book covering the Stolle family tree, that many Stolle family members have. It was written by an Edward J. Stolle, and I'm very grateful that he put this together.

9 20th century just getting started information: A lot of this is common knowledge among history buffs, but it can also be found at www.decades.com, a comprehensive day-by-day look at history.

10 Interview with Ron and Ken Pendery in June 2003. They admit that they can't be sure that this is true, but they seem fairly certain that it is.

11 Speculation from several relatives. Cathy Chasteen always heard that it was a cousin. Ralph's daughter Gail believes that she had heard that it was a Zeigler, a relative on the side of Ralph's grandmother, Anna Stolle, born Anna Zeigler. After

chasing down who it wasn't—the brother of Ralph's best friend—I'm left thinking that he probably did see one of his cousins, a Zeigler, drown.

12 Carolina Street: On-site viewing. In June of 2003, Ron Pendery was nice enough to drive me around a neighborhood in Fort Thomas, where I saw the house on Carolina Street. It's still standing, and the house while in some disrepair was in the midst of being renovated; several workers were actively painting and drilling and fixing up the house. Ron gestured to a field and mentioned that a swimming pool had once been there, until Ralph had it filled.

13 Interview with Ken Pendery, June 2003.

14 Interview with Ron Pendery, June 2003.

15 Interview with Ken and Ron Pendery, June 2003.

16 Steel bars representative: The "Stolle" book indicated that Frank did something with metal, and numerous other people knew it was metal, but no specifics came forth until Dr. Lee Beck mentioned it in an email on October 28, 2003.

17 Machine shop, commuting back and forth: "Stolle."

18 On the homefront: Interview with George Henkle, June 2003, who heard this from Ralph himself. Others have also confirmed this; everybody seems to especially remember hearing about the rabbits.

19 Turn whatever they could into a money-making opportunity: Ibid.

20 Broken finger and stock: E-mail from Dr. Lee Beck on October 29, 2003. The fact that he played first-base, Ralph mentions, on the only tape known in existence.

21 Vera Yearout information: Interview with Vera Yearout conducted by her son-in-law during the summer of 2003.

22 Dinner table conversation: Mini-interview with Ed Cranmer, May 2003.

23 His spouse knew virtually nothing: Mentioned by George Henkle, Dick Pope and others.

24 Worried about taxes: Interview with one of Ralph's attorneys and close friends, Nelson Schwab, September 2003.

25 Not to share the business side of life with his wife: Mini-interview with Ed Cranmer, May 2003.

26 Training him to take care of the other children: Interview with George Henkle, July 1, 2003.

27 Resented mother: Ibid.

28 "I milked two cows..." A taped conversation in which Ralph is interviewed, made in 1996, by two unidentified women from Wilmington College. They apparently were interviewing Ralph for their magazine or newsletter.

29 Early morning, collecting eggs, milking cow, shoveling manure: Interview with Ralph's daughters, Sandra, Gail and Mary Jo, on September 17, 2003.

30 Part of Ralph's duties was to cook: Interview with George Henkle, June 2003.

31 It charmed everybody: Interview with numerous employees, friends, colleagues and family members.

32 Blood pudding: Conversation with Ed Stolle, May 2003.

33 Crossing Ralph: Interview with George Henkle, June 2003.

34 Public speaking: Interview with Dick Pope, April 2003, though confirmed with various other sources.

35 Attitude during and after meetings: Interview with Dick Pope, April 2003.

36 Joseph Stolle information: "Stolle."

37 Joseph's son, Frank: Ibid.

38 John Stolle's birth, Joseph's death: Ibid.

39 John, seven children; Frank, six children: Ibid.

40 Information on John and Anna: Ibid.

41 The demands on Ralph, crowded house: Interview with George Henkle.

42 Close bond: Ibid.

43 Tiny, tough, wooden spoon: Ibid.

44 Free time, milking cow: Ibid.

45 From his grandmother, became interested in farming: Interview with Dr. Lee Beck on July 25, 2003.

46 Stolle Grove: The Kentucky Post, May 21, 1909. Stolle Grove is mentioned in an obituary for the wife of John Stolle. The Stolle Grove information is correct, but the person who died in 1909 was not the wife of John Stolle. Anna Stolle wouldn't die until 1934. The obituary doesn't mention the woman's first name, just that "Mrs. Stolle, wife of John Stolle… died yesterday after a lingering illness. The decedent was 86 years old. My theory is that they were writing about John Stolle's mother, Christine, who I don't have a birth and death date for, but easily could have been 86 in 1909.

47 John Stolle information: "Stolle" book.

48 Mortgage, taxes: Undated interview with George Henkle, by Cathy Chasteen.

49 Stolle & Sons: "Stolle" book.

50 Designing houses: Kathleen Stolle in September 2003 put me onto this lead, and then Dr. Lee Beck confirmed it and added some details. He thinks that Ralph was paid, but he's not positive.

51 Reuben information and splinters anecdote: Interview with Reuben's son, Roger Aschenbach, on October 13, 2003.

52 Will Dorothea be My Aunt: *The Highlander*, the yearbook for Highland High School in Ft. Thomas, KY.

53 General search on the web for Hi-Y clubs; and the YMCA's official web site discusses the history of Hi-Y clubs.

54 A 150 year document talking about the YMCA in Canada. Still, it's relevant to America. The web page is http://www.ymca.ca/eng_150_recreat.htm

55 "History of Ralph J. Stolle." My guess is that it was written in the 1960s, simply because it doesn't discuss the 1970s or beyond. The two page typed document was probably dictated to a secretary, and it's the closet thing to a journal that Ralph left behind.

56 Ostracized: I'm making a guess here, based on what information I have from Ralph's daughters, and from the minimal information that Ralph left behind.

57 Walking to and from school: Mary Jo Cropper, at the September 17, 2003 interview.

58 Chemical engineering student: "History of Ralph J. Stolle."

59 Acetone: September 1995 fact sheet on acetone from the Agency for Toxic Substances and Diseases Registry, confirming what it is; confirming where it was made, consult, *Indiana: A New Historical Guide* by Robert M. Taylor, Jr., Errol Wayne Stevens, Mary Ann Ponder, and Paul Brockman. Published in 1989 by the Indiana Historical Society, Indianapolis, IN.

60 Scientists in the 1920s: A web site by Wayne Pakfo, who has a comprehensive chemical engineering web site with its own historical timeline. He also happens to be a chemical engineer with Procter & Gamble.
http://www1.cems.umn.edu/orgs/aiche/archive/history/h_time.html

61 Universities adding chemical engineering departments: A thorough search through the web at various university web sites confirms this. England, which was slow to jump on the chemical engineering bandwagon, nonetheless had four colleges in the 1920s with chemical engineering classes, and countries like China, Czechoslovakia and even Iran were adding chemical engineering classes and departments.

62 MIT reference: Confirmed at both the University of Cincinnati web site and the web site for MIT (Massachusetts Institute of Technology).

63 "Unfortunately, my Dad went financially broke…" The two-page history of Ralph Stolle.

64 Night school: George Henkle remembers hearing this, for one.

65 Ralph's exhausting schedule: The two-page history of Ralph J. Stolle.

66 1920s references: Historical books, including "The 20th Century: An Illustrated History of Our Lives and Times," edited by Lorraine Glennon (JG Press, 2000).

Chapter Two

ENDNOTES

1 Newspaper ad info: Email from Dr. Beck on October 28, 2003.

2 Where the $250 came from: Interviews with Ralph's family, Dick Pope, George Henkle and others.

3 Hudson reference: I can't be sure of this. It's a guess. Vera Yearout, a cousin of Ralph's, remembers that Frank and Kunigunde had a big Hudson, but she didn't move in with them until 1928. Still, the Hudson was popular in the early 1920s, and so I think it's a good guess.

4 1923: This is the date that virtually every Stolle Corporation document, when discussing the company's history, lists as the company's founding. So it's a safe guess that this is when Ralph moved downtown. We know that he started the business in his garage in 1922, because he writes that in a history of the company, and it's been well-established throughout the family that he bought his own business when he was 18.

5 Sorry for his dad, mom on warpath, employing dad: Email from Dr. Lee Beck on October 28, 2003.

6 Kunigunde believes husband owns business: Interview with Lee Beck in August, 2003, and he confirms this in an email October 28, 2003.

7 Frank's paycheck, lack of one for Ralph: Email from Dr. Lee Beck, October 28, 2003.

8 Stolle Steel & Iron Company: the name is mentioned in a couple old records of Ralph's, including a 1928 tax record, referring back to the previous year.

9 Baseball web site:
http://www.ballparksofbaseball.com/past/CrosleyField.htm

10 brick buildings, slaughterhouses, the smell: Interview with Dick Pope on October 27, 2003.

11 The Cincinnati Enquirer, "City's Products tell its story." December 30, 1999.

12 The Cincinnati Enquirer, "The River Resilient, but still in Danger." October 23, 2000. Jeanne Ison, public relations director for the Ohio River Valley Water Sanitation Commission, offers the 720 dead horse analogy.

13 Interview with Dick Pope on October 27, 2003.

14 Ralph's car: E-mail anecdotes from Dr. Lee Beck on October 27, 2003.

15 Street car anecdote: Email anecdote from Dr. Lee Beck on October 29, 2003.

16 University of Cincinnati, Applied College of Sciences' web site:
http://www.uc.edu/info-services/ocashist.htm

17 Ibid.

18 University of Cincinnati press release dated November 20, 2002, titled Nov. 20 Anniversary to Celebrate College of Applied Science's Deep Roots in Cincinnati.

19 Ibid.

20 Ohio Mechanics Institute college course catalog, dated 1931-1932, the closest thing the university has to the period when Ralph was attending OMI, from 1921 to 1925. It's an interesting document; in many ways it resembles college course catalogs of modern-times until you reach a jolting reminder of how times have fortunately changed, with a sentence that reads on page 13, "Assurance of good moral character is also required. Negroes are not admitted."

21 Ibid.

22 Ibid. Page 15.

23 "History of the Stolle Corporation and the Ralph J. Stolle Companies."

24 Two page document from the Stolle Corporation entitled, "History: Ralph J. Stolle."

25 Frank believed it was his business, Kunigunde did, too: interview with Lee Beck on July 25, 2003.

26 In Ralph's old records from the 1920s and 1930s is a ripped envelope with the address of the Dayton Plating Company printed on it. Frustratingly, nothing was in the envelope.

27 Ralph, president and Aschenbach, secretary: I found this on a gift of two shares of stock that Ralph gave his mother-in-law.

28 Email from Dr. Lee Beck on October 28, 2003.

29 "You can't work with that crazy guy…" Undated notes from an interview with Dick Pope, conducted by Ralph's granddaughter, Cathy Chasteen.

30 Aschenbach moves to Sidney info: Interview with Reuben's son, Roger Aschenbach, on October 13, 2003.

31 Until their deaths in the 1990s: Ibid.

32 Brief interview with Sandy Perry, Ralph's daughter, on September 25, 2003. She has the information from a letter that Florence, later in life, wrote to Dorothy.

33 Historical web page from the Centers for Disease Control and Prevention.

34 Ibid. And Mary Jo Cropper showed me the passage that Robert had written in his diary.

35 Interview with Mary Jo Cropper in July, 2003.

36 Conversation with Cathy Chasteen in June, 2003.

37 Florence and Dorothy, struggling financially: Interview with Mary Jo Cropper in July 2003.

38 Document that I found in Florence Judkins handwriting to the Citizens Bank and Trust Company dated September 28, 1927. Because Florence was living in a family's attic at the time, it seems a logical conclusion to make that she lost her house to the bank, and that she was able to get it back, thanks to Ralph.

39 They never got along: According to Ralph's daughter, Mary Jo Cropper.

40 Honeymoon activities: Ibid.

41 Ralph's new house: Email interview with Dr. Lee Beck on October 29, 2003.

42 Incorporation date, name of company: Mentioned in some tax records of the time.

43 Address of renamed company: Several undated documents, written by unidentified people, that relate the history of the Stolle Corporation indicate that the business Ralph purchased in 1923 was the one on McMicken. But tax records and anecdotal evidence show that Ralph had his earlier business at another street address, and that McMicken was the second place he opened up shop.

44 Interview with Vera Schwitzer, a cousin of Ralph's, conducted during the summer of 2003 by Vera's son, Bob.

45 Dr. Clay Crawford: Ibid.

46 The business was sicker than he was: Numerous people, including Dick Pope, George Henkle and Bob Mays.

47 Tax information: Ralph's tax records from the 1928 and 1929.

48 Interview with Bob Mays in May of 2003.

49 Dragging himself to work: Interview with George Henkle, July 2003.

50 Collapse in bed: Ibid.

51 Web page (http://www.umdnj.edu/~ntbcweb/history.htm) from the National Tuberculosis Center.

52 Brautigan's experience with tuberculosis: Interview with Roger Brautigam, October 2, 2003.

53 Deal with God: Interview with Kathleen Stolle, September 2003.

54 Practicing breathing: Interview with George Henkle.

Chapter Three

ENDNOTES

1 Interview with Dick Pope on January 22, 2003, conducted by Ralph's granddaughter, Cathy Chasteen.

2 Web page from the York Street Café, which is now housed in the building where the Newport National Bank used to be:
http://www.yorkstonline.com/history.html

3 He simply didn't have… running. Interview with Dick Pope on January 22, 2003.

4 That the jumping was a myth is mentioned in the 2003 movie Seabiscuit, which I happened to see during the course of writing this book.

5 Interview with Campbell Graf in June, 2003.

6 …when the stock market crashed… "It was frightful." Two interviews with Harold Hall in April, 2003.

7 Interview with Ralph's daughters, September 17, 2003.

8 1930. In the interview with Dick Pope by Cathy Chasteen, Pope suggests that this incident occurred in "1931, somewhere in that range." I believe it was 1930, which is certainly in that range, because by 1931, he was working on the Empire State Building, and 1930 was simply a very bad year for Ralph.

9 He laid the keys… "I can't do anything." Interview with Dick Pope, by Cathy Chasteen, January 22, 2003.

10 "You can either give me an extension…" Email from Dr. Lee Beck on October 28, 2003.

11 "I don't want the keys… your company going." Ibid.

12 The Kentucky Post, October 8, 1930, "School Board Content in Ft. Thomas," page one.

13 Frank won… were reported. The Kentucky Post, November 9, 1930. "Blond Bandit Captured" and "Campbell-Co is Swept By Democrats."

14 The web site for the Aluminum Anodizers Council is www.anodizing.org.

15 An article from the Products Finishing magazine web site, from an article titled "Anodizing Aluminum" by J.G. Heckler, Jr.

16 Frigidaire and Lowell Gray: Undated interview with Dick Pope conducted by Cathy Chasteen.

17 Close had only been with the company… politics. "From Monopoly to Competition: The Transformations of Alcoa, 1888-1986" by George David Smith, page 336 and 263.

18 Interviews with numerous people.

19 "God… peddler." "From Monopoly to Competition."

20 A bit of a conclusion based on what I know of Ralph and from statements like the one in the September 22, 2003 issue of The Cincinnati Business Courier, in which CEO Dan Cunningham, President/CEO of Long-Stanton Manufacturing Co. Inc., says that the best advice he ever received came from Ralph: "Ralph Stolle told me once that when you're in a business situation, ask this question, 'Tell me exactly what it is you want me to do.'" 21 From Monopoly to Competition: The Transformations of Alcoa, 1888-1986 by George David Smith, page 336.

22 Ibid.

23 Ibid. Page 263.

24 A bit of a conclusion based on what I know of Ralph and from statements like the one in the September 22, 2003 issue of The Cincinnati Business Courier, in which CEO Dan Cunningham, President/CEO of Long-Stanton Manufacturing

Co. Inc., says that the best advice he ever received came from Ralph: "Ralph Stolle told me once that when you're in a business situation, ask this question, 'Tell me exactly what it is you want me to do.'"

25 Wonders of the World Databank, a web site put up by PBS. The clunky web site address is:
http://www.pbs.org/wgbh/buildingbig/wonder/structure/empire_state.html

26 Letter from Frederick J. Close to Dick Pope, dated January 14, 1984. Close was declining an invitation to Ralph's 80th birthday party. Ralph was still in excellent health; unfortunately, Close was not. He also called Ralph, "a great family man, a marvelous industrialist, a good philanthropist and a creative innovator."

27 Interview with Ron and Ken Pendery in June 2003.

28 Agreement by anybody who knew Frank Stolle.

29 Aschenbach had been doing some work… hopped on the train… Undated interview with Dick Pope, talking to Cathy Chasteen.

30 Met with Robert James Anderson: Interview with Tom Anderson on September 26, 2003.

31 Maytag technology, never-crush ringer: Ibid.

32 …by April 15, 1934… three-percent discount." As noted, this all comes from the actual contract, which I found in an old accordion file of Ralph's papers that were lent to me for this book.

33 "This is strictly secondhand… another one." Interview with Tom Anderson on September 26, 2003.

34 $2,500 loan… Ibid.

35 …a punishing heat… wandered away from home. Information collected from the archives of the Shelby County Historical Society, in Sidney, Ohio.

36 "The Sidney Daily News," August 25, 1978, "Stolle came to Sidney in Depression year 1934."

37 Interview with Zwiesler's daughter, Norma Counts, on September 26, 2003.

38 Ibid. That he was hauling aluminum is my conclusion.

39 Ibid.

40 Ibid.

41 Ibid.

42 Interview with Tom Anderson.

43 Ibid.

44 Undated interview by Cathy Chasteen with Dick Pope. Obviously, since Dick Pope is sharing a story that was told to him, we have no idea if the dialogue is exactly how the exchange went, but I think we can assume it was close.

45 There are so many different years out there, in various accounts, regarding the fire, but I contacted the Shelby County Historical Society, which is in Sidney, Ohio, and according to their records, the fire was in 1939.

46 "Dad and him were watering it down…" Interview with Norma Counts on September 26, 2003.

47 Memories of 1939: Interview with Maxine Pence in July 2003.

48 Robert Anderson's fate: Interview with Tom Anderson on September 26, 2003.

49 "Ralph always said…" Ibid.

Chapter 4

FOOTNOTES

* George Henkle, it should be noted, was 92 years old when he made this comment and just eight years younger than Ralph. But he is a young 92.

* Based on one conversation I had with somebody in the position to know and whom I believe let the number slip, my guess is that the figure was $10 million.

ENDNOTES

1 "We would spend…" Interview with Mary Jo Cropper in August, 2003. Throughout this chapter are anecdotes from Mary Jo, most of which came from this interview.

2 "He was moving too fast…" Interview with George Henkle, July 2003.

3 Interview with Rita Hauck in April, 2003.

4 Seersucker information: Cigar Aficionado magazine, "American Dreams," the November/December 1998 issue.

5 Seersucker information: Ibid.

6 Cathy Chasteen, interview with Mary Jo Cropper in August 2003.

7 Mary Jo Cropper interview.

8 Ibid.

9 "Upon graduating from high school… that car!" anecdote shared by Mary Jo Cropper in an email dated September 21, 2003.

10 Consulted MapQuest.com.

11 According to MapQuest.

12 Interview with Mary Jo.

13 Staying awake: Interview with Bob Mays. The "son he never had" is my own conclusion.

14 Half mile of cars: Interview with George Henkle in July 2003.

15 Interview with Ron and Ken Pendery in June 2003.

16 Date and Dobbling funeral home reference: Unreferenced newspaper clipping provided by the Cincinnati Historical Library. Frank's obituary, apparently from The Cincinnati Enquirer, dated October 17, 1952.

17 Family members gathered: Interview with Ron and Ken Pendery, June 2003.

18 Ohio Department of Natural Resources web page: http://www.dnr.state.oh.us/parks/parks/indianlk.htm

19 Interview with Mary Jo Cropper.

20 Why Ralph didn't move the family to Sidney: Ibid.

21 Interview with George Henkle.

22 Description of home: Largely my own observations from a tour of the home given to me by grandchildren Ed Cranmer and Cathy Chasteen.

23 Interview with Mary Jo Cropper.

24 20 minutes: Mentioned by several people, including Mary Jo Cropper.

25 Ibid.

26 Mentioned by Dick Pope and others.

27 Weekend parties quote: George Henkle interview.

28 Polio information: came largely from a web site, put up by Edmund J. Sass, Ed. D., author of Polio's Legacy: An Oral History.

29 Daughter kidnapped: Interview with Ralph's daughters, September 17, 2003.

30 Newspaper provided to be the Cincinnati Historical Library. The name of the paper isn't attached to the article, but it was dated September 10, 1937, and the article is titled, "Strike Call Fizzle." Ralph was quoted extensively in the article, and he summed up his feelings by saying: "We have done everything possible to prevent strife and our employees have unanimously expressed their satisfaction. We feel that the strike order was unfair and unwarranted. We shall continue to operate our plant." Indeed, they did.

31 Interview with Dick Pope, August 2003.

32 Ibid.

33 Mentioned in the aforementioned September 10, 1937 article.

34 Interview with Dick Pope, August 2003.

35 McCarthyism: Ibid.

36 Actual paper poster that I have, from the files of Ralph J. Stolle.

37 Pattern of violence, Forrest Blankenship: From the archives of the Shelby County Historical Society. Basically, they have recorded highlights of each year of Sidney's history, as far back as the daily paper goes, and the Stolle union troubles were one of the year's more colorful news events.

38 An enormous bomb: Ibid.

39 Rita Hauck information: Interview in April 2003

40 Millionaire fixing breakfast anecdote: Interview with Weldon Oakley at a retirement luncheon in July 2003.

Chapter 5

FOOTNOTES

*The first time church key was used in print was 1951, but there is anecdotal evidence that suggests people had been using the terms since the late 1930s.

ENDNOTES

1 Dayton Daily News, March 14, 1988, "Pop a Top, toast one of our own."

2 Undated, unreferenced newspaper clipping, apparently cut out by the Stolle family or an associate. It's an Ermal Fraze obituary, probably from the Dayton Daily News, in the year 1989. And the story is repeated in numerous texts about the invention.

3 Dayton Daily News, March 14, 1988.

4 Letter received from Elton Kaminski on October 20, 2003, answering various questions of mine, and which described much of the chronology of the Stolle Corporation's involvement with the easy open pull tab.

5 A web site put up by Cummings Properties, which owns the building that the former United Shoe Machinery Corporation once resided in. The web site also mentions that the United Shoe Machinery Corporation developed the easy open pull tab.

6 Letter from Elton Kaminski on October 20.

7 One of the Alcoa branches… busy with other projects. Ibid.

8 It was then… Ibid.

9 Ibid.

10 Ibid.

11 Ibid.

12 Fraze accepted… Elton Kaminski's letter on October 20, 2003.

13 Fraze patented his work… Ibid.

14 They controlled the licensing of Fraze's patent. Elton Kaminski's email on October 24, 2003. March 14, 1988, Dayton Daily News also mentions that Fraze had a patent.

15 From Monopoly to Competition: The Transformations of Alcoa, 1888-1986 by George David Smith (Cambridge University Press), 1988.

16 Ibid.

17 Ibid.

18 Ibid.

19 Ibid.

20 Ibid.

21 Ibid.

22 Ibid.
23 Ibid.
24 Ibid.
25 Ibid.
26 Interview with Dick Pope, April 2003.
27 Ibid.
28 Ibid.
29 Ibid.
30 "Sure, I can." Ibid.
31 Letter from Kaminski, October 20, 2003.
32 Fraze wasn't there… Ibid.
33 "allowed high speed indexing movements through the tooling." Email from Kaminski, October 24, 2003.
34 Email from Kaminski, October 24, 2003. "Ralph understood perfectly what I was explaining to him," wrote Kaminski.
35 Kaminski letter, October 20, 2003.
36 Meanwhile… look like. Ibid.
37 Ibid.
38 Ibid.
39 "The rush… was on." Ibid.
40 An assumption. That's simply what Ralph did. He sent his plane out to pick up the people he did business with.
41 He traveled back to Lebanon… fragments of the discussion. Interview with Nelson Schwab on September 18, 2003.
42 Ibid.
43 The letter, which is framed, was lent to me by Dick Pope, who keeps it in his house as a souvenir of those heady times.
44 Interview with Dick Pope in August, 2003, and my own conclusions from talking to numerous people at a Stolle Corporation retirement luncheon.
45 Interview in April, 2003, with Bob Mays.
46 Interview with Dick Pope in August, 2003.
47 Notes from Kaminski.
48 Interview with Dick Pope in August, 2003.
49 Notes from Kaminski.
50 Notes from Kaminski.
51 Interview with Tom Anderson, on September 26, 2003.
52 Both his father… maintenance and engineering. Notes and letter from Kaminski, October 20, 2003.
53 Interview with Roger Brautigam, September 26, 2003.

54 Bill Littlefield remembers Charlie Williams… lids of beer cans. Interview with Bill Littlefield, September 24, 2003.

55 Ibid.

56 "the situation was urgent…"… extra overtime income. Ibid.

57 "In the tool and dye business… You're into it." Interview with Roger Brautigam, September 2003.

58 Sandy's quote. Mary's Jo's. Interview with Ralph Stolle's daughters, on September 17, 2003.

59 Slim had been making deliveries… making a beeline to the Stolle Corporation." Interview with Slim in May, 2003.

60 Interview with Dick Pope, April 2003.

61 "Their plain can ends… convincing the canmakers." Notes from Kaminski.

62 Anecdote involving Ralph dipping can ends in the paraffin, came from Kaminski's notes.

63 The Sidney Daily News, August 25, 1978, "Stolle Develops Flip Up Beverage Can Lid," part of a retrospective of stories that the newspaper did on the 25th anniversary of the easy open pull tab invention.

64 From Monopoly to Competition, page 341.

65 The Stolle Corporation, 1923-1964, a one-page document detailing the company's history.

66 The Stolle Corporation, 1923-1964.

67 List of companies: "Stolle," the family reference guide.

68 Definition comes from a web page designed by Alcoa: http://www.alcoa.com/alumina/en/info_page/alumina_defined.asp

69 Monopoly and Competition, page 342.

70 Web page from AZoM, dedicated to being a source for engineers and the design community: http://www.azom.com/details.asp?ArticleID=1483

71 Undated, unreferenced newspaper clipping that was clipped out and saved by Ralph, or one of his friends or family members. The convicts were suing for more than $4 million.

72 Invention by Design; How Engineers Get from Thought to Thing by Henry Petroski, page 102-103.

73 Letter from Kaminski, on October 20, 2003.

74 Melanoma anecdote. I found this at a web site from the Urban Legends Reference Page, an excellent, all-comprehensive web site dedicated to ferreting out urban legends: http://www.snopes.com/business/redeem/pulltabs.asp

75 Ronald McDonald House Pop Tab Collection Program web site: http://www.rmhc.com/kids/poptab_collection/

76 Photo of Ermal Fraze's mausoleum, at the www.findagrave.com web site. Yes, there is such a web site.

Chapter 6

ENDNOTES

1 My own conclusions, drawn from two extensive discussions about Mike Seaving with George Henkle in July of 2003 and Dick Pope in August 2003.

2 From the moment… white Ford Thunderbird. Interview with Dick Pope, August 2003.

3 And then there were times… into the clink. Interview with George Henkle, July of 2003.

4 Description of Seaving. Interview with Dick Pope, August 2003.

5 But he was no pushover… six feet. Ibid.

6 Information from both Pope and Henkle.

7 …eccentric character… box from the pocket. Interview with Dick Pope, August 2003.

8 "Ralph knew…" Interview with Maxine Pence, July 2003.

9 …a 310 Cessna… grass runway. Memories from Slim, in an interview in May 2003.

10 He would spend nights… about the cost. Ibid.

11 Article dated December 11, 1950. It was a clipping provided to me by the Cincinnati Historical Library; there is no name of the paper attached to the clipping, but I suspect it's from The Cincinnati Enquirer or the Cincinnati Post.

12 The parties… nametags. Interview with George Henkle, July 2003.

13 Renting a bus: Interview with George Henkle, July 2003.

14 80 employees' children to college: Numerous people have told me this, but in particular Campbell Graf, who mentioned it in May 2003.

15 When that person was rich: Ibid.

16 Starting a restaurant: Conversation at retirement luncheon.

17 Forty-two farms: Picked up this bit of information when interviewing Ken and Ron Pendery in June 2003, but I'm unsure who told me this.

18 Dan Phelps: Interview with Slim Buchanan in July 2003.

19 Art Middleton info: Interview with Art Middleton in July 2003.

20 Delivering food: Interview with Slim Buchanan in May 2003.

21 Crutches anecdote: Dalton and Linda Messersmith interview in July 2003.

22 Interview with Dick Pope in August 2003. And other employees mentioned that they were laid-off at one time or another.

23 Richard Borchers: All of his stories come from a letter that he was nice enough to send to me after the Stolle luncheon.

Chapter 7

ENDNOTES

1 Interview with Dr. Lee Beck on July 25, 2003. And let's just save a lot of time by saying that anything in this chapter involving Dr. Lee Beck, came from our interview on July 25 or a subsequent email exchange on July 28.

2 Ibid.

3 Email from Dr. Beck on October 27, 2003.

4 Coffee at desk: Interview in early July at the Stolle retirement luncheon. I don't know why this employee didn't want her name used. Does she think her pension might be revoked?

5 Filet mignon: Interview with Janet Aikin at the Stolle retirement luncheon in July 2003. Now, see she used her name.

6 Even-tempered: Letter from Harold Hall to Cathy Chasteen, dated December 28, 2002.

7 A copy of a letter, written to nobody in particular, dated July 22, 1990. The letter isn't signed by Ralph, but it appears to have at least been dictated by him.

8 Community leaders: Mentioned at the Ralph J. Stolle Countryside YMCA web site.

9 The Cincinnati Enquirer, July 18, 1999, "The Reluctant Hero."

10 "...probably seven on a 10 scale." E-mail from Neil Armstrong, October 2003.

11 Golf, pheasants, cooking: Ibid.

12 Moon launch: Mary Jo Cropper interview in July 2003.

13 Armstrong's visits: Interview with Slim Buchanan in June 2003.

14 Rhodes' appearance: The Western Star, County YMCA Now Official. November 8, 1978. The article describes the dedication; the public got to go inside two months earlier, however.

15 Information about the YMCA: the Ralph J. Countryside YMCA web site

16 Inside the YMCA: Ibid.

17 Ralph and his colleagues: Interview with George Henkle, July 2003.

Chapter 8

ENDNOTES

1 Description of old Stolle haunts: Slim Buchanan drove me around to show me where Ralph used to work.

2 Miami University: Mary Jo Cropper, interview July 2003.

3 Fritz Close's year of death: recollection from Dick Pope.

4 Ermal Fraze's year of death: Photo of his gravesite, which can be located at www.findagrave.com

5 Reuben's date of death: I managed to find the information from his death certificate on the Internet.

6 Pauline Yergin's year of death: Recollection from several people, and the Cincinnati Enquirer's obituary.

7 Literally wore her out: Recollection from several people, including Lee Beck and George and Lois Henkle.

8 Mutual friend, all information and quotes from Kathleen: Interview with Kathleen Porter Stolle in September, 2003.

9 Mentioned by several people, including Cathy Chasteen and George Henckle.

10 Disdain for Cadillac: Mentioned by several people; he was almost legendary for his dislike of the car.

11 Ignored journalist: Mentioned by Ed Cranmer, Ralph's grandson.

12 Ralph J. Stolle Countryside YMCA web site.

13 Canadian Broadcast Corporation news story on the Concorde from April 10, 2003.

14 Description of YMCA: Ralph J. Stolle Countryside YMCA web site.

15 Buying land around Wilmington: Interview with Campbell Graf in May 2003.

16 Campbell Graf's thoughts on Ralph: Ibid, which I think is Latin for "ditto."

17 $1 million: Mentioned a few times in various Wilmington business school literature.

18 Cincinnati Children's Hospital: Interview with Lee Beck in August 2003.

19 Millions and millions: Ibid, and the opinions of others I've interviewed.

20 Wilmington College piece: recollection from Dick Pope, October 27, 2003.

21 Taped interview of the two women from Wilmington interviewing Ralph, provided by Mary Jo Cropper. At one point on the tape, Dick Pope asks Ralph about his childhood, but one of the women interrupts, saying, "Oh, we have that on another tape." But that tape is apparently lost for good, reports Mary Jo, sounding understandably disappointed. I was pretty crushed, too.

22 Recollection from a letter from Danny Lewis, sent to Ralph around the time of Ralph's 90th birthday.

23 Oyster Stolle recipe: Ibid.

24 Neil Armstrong's memory of hasenpfeffer: Armstrong's e-mail.

Epilogue

ENDNOTES

1 Sharp pain, "This is it…": Interview with Kathleen Stolle, September 2003.

2 Prostate cancer: Ibid.

3 His obituaries in the Cincinnati Enquirer, New York Times, USA Today…

4 Stolle Milk Biologics press material.

5 Milk protein concentrate into products: Interview with Lee Beck, August 2003.

6 Thomas Edison: Ralph comments on him, during the tape with the Wilmington ladies.

7 Portrait: I saw it for myself; you can't miss it.

About the Author

Geoff Williams has been widely published in magazines, including *LIFE, Ladies' Home Journal* and *Entertainment Weekly.* Most frequently, his work can be found in two national magazines, *Entrepreneur*, in which he is a frequent contributor, and *BabyTalk*, where his monthly column explores the lighter side of fatherhood. He lives in Loveland Ohio, with his wife Susan, their two daughters, Isabelle and Lorelei, four cats, an aquarium full of fish and one dog.